Hope Jahren is an award-winning scientist who has been pursuing independent research in paleobiology since 1996. Recognised by *Time* magazine in 2016 as one of the 100 most influential people in the world, she is the recipient of three Fulbright Awards and served as a tenured professor at the University of Hawaii in Honolulu from 2008 to 2016, where she built the Isotope Geobiology Laboratories. She currently holds the J. Tuzo Wilson Professorship at the University of Oslo as an elected member of the Norwegian Academy of Science and Letters.

thestoryofmore.com
hopejahrensurecanwrite.com
jahrenlab.com

'Hope Jahren is the voice that science has been waiting for'
Nature

'A superb account of the deadly struggle between humanity and what may prove the only life-bearing planet within ten light years, written in a brilliantly sardonic and conversational style'
E. O. Wilson

'Hope Jahren asks the central question of our time: how can we learn to live on a finite planet? *The Story of More* is thoughtful, informative and – above all – essential'
Elizabeth Kolbert, author of *The Sixth Extinction*

ALSO BY HOPE JAHREN

Lab Girl

The
Story
of
More

How we got to
climate change
and where to go
from here

Hope Jahren

FLEET
2020

FLEET

First published in the United States in 2020 by Vintage Books,
a division of Penguin Random House LLC
First published in Great Britain in 2020 by Fleet

1 3 5 7 9 10 8 6 4 2

A CIP catalogue record for this book
is available from the British Library.

ISBN 978-0-7088-9898-7

Book design by Nicholas Alguire
Printed and bound in Great Britain by Clays Ltd, Elcograf S.p.A.

Papers used by Fleet are from well-managed forests
and other responsible sources.

Fleet
An imprint of
Little, Brown Book Group
Carmelite House
50 Victoria Embankment
London EC4Y 0DZ

An Hachette UK Company
www.hachette.co.uk

www.littlebrown.co.uk

For my mother
and
because of my father

Contents

Part One

LIFE

The universe is change; our life is what our thoughts make it.
   —Marcus Aurelius Antoninus (A.D. 121–180)

1

Our Story Begins

The sun and solar energy. What a source of power! I hope we don't have to wait until oil and coal run out before we tackle that.
——Thomas Edison to Henry Ford
and Harvey Firestone (1931)

Important men have been arguing about global change since before I was born.

Almost ninety years ago, the guy who invented the light bulb urged renewable energy on the guy who invented the car and the guy who invented the tire. I imagine they nodded politely, finished their drinks, and went straight back to motorizing the planet. During the decades that followed, the Ford Motor Company manufactured and sold more than three hundred million motor vehicles that burned upward of ten billion barrels of oil and required a minimum of 1.2 billion tires, also partially made from oil.

But that's not all. Back in 1969, the Norwegian explorer

Bernt Balchen noticed a thinning trend in the ice that covered the North Pole. He warned his colleagues that the Arctic Ocean was melting into an open sea and that this could change weather patterns such that farming would become impossible in North America ten to twenty years hence. *The New York Times* picked up the story, and Balchen was promptly shouted down by Walter Whittmann of the U.S. Navy, who had seen no evidence of thinning during his monthly airplane flights over the pole.

As is the case with most scientists most of the time, Balchen was both right *and* wrong in his claims. By 1999, the submarines that had been cruising the Arctic Ocean since the 1950s could clearly see that polar sea ice had thinned drastically during the twentieth century—thinned by almost half. Nevertheless, it's been fifty years since Balchen graced the pages of the *Times* and American agriculture has yet to feel the full effect of any melting. Which, technically, means that Whittmann was also both wrong and right.

We shouldn't be surprised when scientists are wrong. All human beings are a lot better at describing what is happening than at predicting what will happen. Somewhere along the way, however, we began to hope that scientists were different—that they could be right all the time. And because they're not, we kind of stopped listening. By now we're quite practiced at not listening to things scientists say over and over again.

For example, giving up fossil fuels is not a new suggestion. Starting in 1956, a geologist named M. King Hubbert who worked for Shell Oil started writing passionately about America's need to embrace nuclear energy before our "inevitable exhaustion of fossil fuels." Hubbert believed that mining uranium from the bedrock of Colorado was more sustainable than burning oil and coal, which he reckoned would hit peak production by the years 2000 and 2150, respectively. He was both wrong and right.

Let's go back to 1969 for a moment, back when Balchen was fighting with Whittmann and Hubbert was still on his soapbox. I don't remember 1969 personally, but, like every year, it was full of beginnings and endings, problems and solutions, equal to any that had gone before or have come since.

Most of the trees you see out your window were barely seeds in 1969. Wal-Mart Stores, Inc., was incorporated in 1969 and has since become the world's largest private employer. *Sesame Street* premiered in 1969 and went on to teach millions of children how to count and spell. Big things started out as small things, then grew to change the world.

When the polluted Cuyahoga River caught fire in 1969, every single fish between Akron and Cleveland died, and *Time* magazine's coverage led to the creation of the Environmental Protection Agency. That same year, an offshore oil platform disgorged more than one hundred thousand barrels of crude oil onto the beaches of Santa Barbara, California, killing every sea creature in its path and spurring the organization of Earth Day, which is now observed around the world.

Way up north, in Mower County, Minnesota, my parents weren't paying attention, for I was one of the ten million babies born on September 27, 1969, and the last of their four children. The world would be different for this baby, my parents promised each other, and they made the ancient vow that all mothers and fathers make in the euphoria that follows a happy birth.

I would have all the love that my father could give and all

the love that my mother should have been given. She will grow up free, my mother resolved—free from hunger and from the shame of being taken by the county. My father, for his part, looked forward to a century of technologies that would save us all from sickness and want. Like the millions of couples who had come before them, and would come after, they looked at the world that they lived in and thought about the one that they wanted. Then my parents turned to each other, in love, and they named me Hope. And they were both right and wrong at the same time.

Forty years later, in 2009, my department chair called me into his office and asked me to teach a class on climate change. I groaned and slouched down into my chair. Convincing people to examine their energy use is like trying to get them to quit smoking or to eat more healthfully: they already know that they should do it, but there's a billion-dollar industry working round-the-clock, inventing new ways to make sure that they don't. I also couldn't help thinking about Edison and Ford and Firestone and Balchen and Whittmann and Hubbert and Sagan and Gore and all the important men who had already tried to raise the subject and, quite frankly, wouldn't have thought much of a lab girl like me. I thought about the car I'd driven to work that morning, and how it leaked oil terribly, and wondered what business I had in telling anybody anything.

I left my chair's office and went back to my lab, where I conferred sulkily with my coworker Bill. I asked him, after detailing the futility of it all, why I should even try. He listened patiently until I was finished and then gave me his usual pep talk: "Because it's your job. *Shut up and go do your job.*"

Bill is the exception to many rules, not the least of which is that he's right far more often than he's wrong. As usual, he had a point. I was making too much out of this. I would go do my job,

I decided, by taking my boss's orders at face value. I sat down at my desk, turned on my computer, and began to research *change*. Over the next several years, I cataloged the data that describes how population has increased, how agriculture has intensified, how energy use has skyrocketed over the last half century. I accessed public databases and downloaded files of numbers and spreadsheets. I searched through the data for patterns across the decades of my own life. I set out to quantify global change in the most concrete and precise terms that I could apprehend, and I learned a lot by doing it.

This research became the foundation of a course that I taught many times. Each week during the semester, I'd pick up the chalk and teach a room full of students the numbers describing how planet Earth had changed since I was a kid back in the 1970s. I taught them what has happened. Not what I think *might* happen. Not what I think *should* happen. I taught them the things that I had taught myself. And as I did my job, I finally began to understand why I was doing it: because only after we see where we are can we duly ask ourselves if this is where we want to be.

Right now, I see the country of my birth moving backward. It has dumped the Paris Agreement, it's close to dismantling the Environmental Protection Agency, and the United States Department of Agriculture is in very bad shape. The United States Department of Energy, which funded my lab for more than a decade to study greenhouse gases, has shut down most of its work on climate change, and NASA is under pressure to do the same. I left the United States in 2016 and moved to Norway because I believe that my laboratory will have more support here and because I am worried about the future of science in America.

All of this has convinced me that it's time to bring global change out of my classroom and into this book. Not because I am a scientist who thinks she's right, but because I am an author who loves both words and numbers and a teacher who has something to say.

So if you'll listen, I'll tell you what happened to my world, to your world—to *our* world. It changed.

2

Who We Are

There was a time when the countless tribes of men, though
*   wide dispersed,*
Oppressed the surface of the deep-bosomed Earth.
                    —Stasinus, *The Cypria* (c. 750 B.C.)

Eighteen hundred years before the birth of Christ, the people of
Mesopotamia were deeply worried that planet Earth could never
provide enough food, water, shelter, and space for the population
they saw increasing all along the Tigris-Euphrates River valley—
which amounted to the whole world, as far as they knew. The
poets of Babylon—Mesopotamia's largest city—told how, soon
after the advent of procreation, "the world teemed, the people
multiplied, the world bellowed like a wild bull." This drove the
gods, deprived of sleep by the din of civilization, to cull human
numbers by way of famine and disease, over and over again.
These poems were written in about 1800 B.C., just as world popu-

lation hit somewhere near one hundred million people. During the one thousand years that followed, global population doubled.

Aristotle believed that politicians should decide "of what numbers and of what sort of people" a country should contain. Because "too large a multitude are incapable" of keeping themselves in good order, Aristotle recommended a suite of regulations on marriage, mainly directed toward controlling women's behavior. Within his treatise *Politics,* he specified that women should not marry too early ("for women who do so are apt to be intemperate") but should not bear children too late, either ("that the children may be equally perfect"). As a bonus, Aristotle itemized a list of instructions for rearing boy children, detailed to a level that makes today's helicopter parents appear positively negligent. Boys must be inured, he insisted, "to cold when they are very little," a sentiment that my parents embraced two millennia later in Minnesota, completely unaware that it had originated, like democracy itself, in ancient Greece.

If the world would just follow this simple set of directions, Aristotle maintained, the population would be appropriately curbed, and Earth would forever be able to supply its inhabitants with the "necessaries of life." He wrote these things in about 330 B.C., just as world population was nearing 250 million. During the one thousand years that followed, global population doubled again.

A medieval abbot named Suger is credited as the founder of Gothic architecture, but his actual goal was to raise money. He wanted to enlarge the Basilica of Saint-Denis in order to accommodate the drastic increase in penitents that was produced during the population explosion of the High Middle Ages. He described a church entrance so crowded that it "forced women to run to the altar on the heads of men as on a pavement," leaving miscellaneous faithful bloodied on the floor. Suger wrote this in 1148, when the world population was about one-half billion people. Within the next five hundred years, global population doubled yet again.

It was Thomas Robert Malthus who really took overpopulation neuroses to the next level with his famous *An Essay on the Principle of Population*. Any and every improvement in food production, he argued, would give rise to an increase in population that would, in turn, reestablish a state of want. Malthus believed that our planet moils within an everlasting state of overpopulation and that every bite we put into our mouths makes things worse. He published his *Principle* in 1798, just as world population was approaching one billion, and since then, entire forests have been sacrificed to pamphlets, columns, treatises, and dissertations discussing his idea. Nevertheless, during the one hundred years that followed, global population doubled again.

Fifty years after Malthus sounded the alarm, John Stuart Mill expanded Malthus's food-security argument to include less tangible goods and services, wringing his hands over the broader economic "penalty attached to over-population." He spelled out his idea of our collective zero-sum game within the *Principles of Political Economy*: "A greater number of people can not, in any given state of civilization, be collectively so well provided for as a smaller." He insisted these things during 1848, just as world population approached 1.5 billion people. And within the next century, the global population doubled again.

I've already told you about M. King Hubbert, who spent his career urging the adoption of nuclear energy while working for Shell Oil. Like a lot of his contemporaries, Hubbert was also obsessed with population control. He didn't have any kids himself, but he knew how many other people should have. Upon hearing that his coworker had gotten married, Hubbert marched up to Mrs. Co-worker and instructed her, "Look, you can have two children, you can have them in series or in parallel, but no more than two." Hubbert presented his ideas to the National Academy of Sciences in 1969, the same year that I was born, which added one small female to what was then a world population of 3.5 billion people. Between then and now—that is, over the course of my lifetime so far—the global population has doubled yet again.

Today, you and I share the world with more than seven billion people.

There's a pattern here that suggests the vocal disapprobation of overpopulation is not sufficient, in and of itself, to stanch population growth. One thing that these great thinkers never explored, however, is the correlation between the status of a woman within her society and the average number of children she bears within her lifetime.

Of the world's ten countries with the lowest gender gap (that is, the *least* difference between male and female health, opportunity, and participation), seven are also considered high-income relative to the rest of the world. Conversely, six of the ten countries with the highest gender gap (that is, those with the *most* difference between male and female health, opportunity, and participation) fall into the low-income categories. Whether this suggests that wealth gives rise to, or is a product of, female health, opportunity, and participation is not clear; it could be a combination of both.

What is clear is that societies that feature a low gender gap are also populated by women who give birth, on average, half as often as women who live in societies with a high gender gap. The average number of births per woman living in the "high gap" countries is close to four, while the number in the "low gap" countries is just under two. It makes sense that the most effective and long-lasting mechanism for curbing global population growth revolves around an elimination of gender inequality.

These data also imply that closing the gender gap around the world would likely result in something near replacement-level fertility: that is, a stable global population that neither increases nor decreases. This best-case scenario means that barring something awful—famine, plague, genocide, or forced control over reproduction—Earth will never again contain fewer than seven billion people. We must learn to live together if we want to live well.

Not all experts abhor the population increase that gave rise to their own birth. An American economist named Henry George gravely opposed Malthusian doctrine on the grounds that it "shelters selfishness from question and from conscience." The way he saw things, the true cause of poverty among the masses was "the rapacity of man" and not the inadequacy of nature. George also posited a hopeful feedback loop between human population and food production: "Give more food, open fuller conditions of life, and the vegetable or animal can but multiply; the man will develop."

Henry George was a humble and pious man; he had known poverty and lived modestly even after success. He believed that intelligence should both inspire sympathy and seek justice. George was ahead of his time in many ways, advocating for public transportation, labor unions, and a half-female Congress during the nineteenth century, but the public loved him. More than a hundred thousand mourners attended his funeral after he died in 1897. It also turns out that Henry George, out of all these important men, was the one who came the closest to being right about population growth.

The doubling of global population that has occurred since the time of Henry George was accompanied by a *tripling* of grain production, a *tripling* in fish production, and a *quadrupling* in meat production. The arrival of more people has indeed coincided with greater food production, just as Henry George believed it would—far above and beyond what was strictly required.

Henry George was also right in that most of the want and suffering that we see in our world today originates not from Earth's inability to provide but from our inability to share, as we'll see so many times in later pages. It is because so many of us consume far beyond our needs that a great many more of us are left with almost nothing.

The more complicated problem, unique to our generation, is that the enormous consumption of food and fuel by just 10 percent of us is actively threatening Earth's ability to produce the basics of life for the other 90 percent. Most political discussions of climate change are predicated on the hope that this may be reversible; the truth is that it may not.

Ultimately, we are endowed with only four resources: the earth, the ocean, the sky, and each other. Because absolutely everything is at stake, it behooves us to begin by thinking clearly and simply. We'll begin where every human story begins. We'll begin with a baby.

3

How We Are

Possibly I know now what the soldier feels when
a bullet crashes through his heart.

—Mark Twain,
upon learning of his daughter's death,
December 24, 1909

I had a friend in college who wasn't afraid of anything. Motorcycle races, cliff hiking, public speaking, camping among bears, foreign languages, late-night confessions: he was first in line, never flinching, never frowning—open and ready for what was ahead and for whatever was beyond that.

When my friend was in his thirties, his wife had a baby boy. Their son was healthy, until he stopped breathing. On day five, the doctors told him that his son had been born with an extra copy of a chromosome, which had severely interfered with the formation of his heart and kidneys. A few weeks later, his son died, and my friend closed. He has spent the rest of his life working to reopen.

wo hundred years ago, losing a child was very common. In 1819, two out of every five children born alive, the world over, died before their fifth birthday. Because family size was much larger than it is today—a woman of that time gave birth, on average, six times during her lifetime—it would have been rare to meet a parent who had not endured this terrible grief at least once, if not twice. By the time I was born in 1969, this great global sorrow had been more than halved: my survival was one of the fortunate four out of five. Any number is still too high, but during the last fifty years, childhood deaths have been reduced drastically: at present, one out of twenty-five babies dies before they turn five. Because families are also smaller now (each woman gives birth, as a global average, about three times during her lifetime), we encounter this great loss and its devastation far less regularly, such that many of us cannot imagine the expectation of a new baby overshadowed by the dread of death.

Of course, in 1819 it was not only the baby that risked death while being born; the mother did as well. Two of my eight great-grandmothers died while giving birth, and my family history recorded next to nothing about the lives (and deaths) of the other six. One hundred years ago, one out of every hundred live births in the United States resulted in a dead mother. Since then, improved medical care in the form of a midwife, a nurse, or a physician with access to sterile medical tools has greatly lessened the risk for both mother and baby during childbirth. Between 1930 and 1980, advances in care brought the United States down to one maternal death for every ten thousand live births, where it has stubbornly remained to this day.

When it comes to deaths associated with childbirth, the United States fares ten times better than the global average. When

I was born in 1969, only about half of the births that occurred on planet Earth took place with the benefit of improved care. By 2013, an increase in medical access of only 20 percent had reduced global maternal deaths by more than half. Today, one in five hundred live births around the world results in the death of the mother.

The history of most nations reveals a dramatic decrease in maternal and child mortality during the last three centuries. If we envision ourselves on a global journey toward eliminating the agony of loss that our grandparents endured, we should rejoice at how very far we have traveled, even while acknowledging the distance ahead. Best of all, we already know the way forward.

What do you have planned for your sixties? I hope to still be riding my bicycle regularly and picture myself coasting downhill through a light blizzard of cottonwood fluff during June 2034. I stop at dusk to lock up outside of Target Field and take my seat just in time to see the Minnesota Twins win handily in eight and a half innings, 17–0 against the Yankees.

I have good reason to make such plans, though my father never did. When he was born in 1923, the life expectancy for a newborn American child was only fifty-eight years. When I was born almost five decades later, that number had increased such that I could expect to live to be seventy-one. My son must plan not only his sixties but his seventies as well: by the time he was born in 2004, American life expectancy had increased to seventy-eight years. Maternal optimism compels me to encourage him to plan beyond eighty, and not without cause, for my own father was a true outlier who lived past his ninety-second birthday. He died of pneumonia, which is an infectious disease of the respiratory system, and while it is a common cause of death in ninety-plus-year-olds, it is otherwise a very unusual way to die in America.

For example, he could have been one of the more than four hundred thousand American men who died while fighting in World War II between the years of 1939 and 1945, but he wasn't. World War II resulted in a terrible loss of life, with perhaps fifteen million soldiers and as many as forty-five million civilians killed worldwide over less than a decade. During the seventy-plus years since, human civilization has not witnessed anything like it. In fact, the number of deaths around the globe each year due to wartime conflicts averages about fifty thousand people, as of the last few years.

Conventional homicide continues to kill many more people than war does: five hundred thousand murders occur each year around the globe. But murder and war, even when put together, pale in comparison with the annual loss of life due to suicide: our world witnessed almost eight hundred thousand suicides during 2016, fifty thousand of them occurring within the borders of America. For all the violence we perpetrate against one another, we inflict much more upon ourselves.

The vast majority of human deaths that occur on planet Earth each year, however, are due simply to illness. Death is known as the great equalizer, so it may not surprise you to hear that people die at about the same rate in every region, be it rich or poor or in between. Every year, the world over, about 1 percent of the population becomes sick and dies. People succumb to very different illnesses, however, that are strongly associated with economic status. In the richer countries, the top two killers are heart disease and stroke, which account for one out of every four deaths. Taken together with cancer, diabetes, and kidney disease, they are responsible for about half of all deaths within the world's wealthiest nations. Each of these diseases is dreadful in its own way, but they have something important in common: none of them are contagious.

Within the poorest regions of the world, the story is very different. People may be dying at the same crude rate, but they suffer and expire at a younger age and from diseases that have been almost eliminated within richer countries through access to clean water, improved sewage systems, vaccination, and anti-biotics. Contagious infections of the lungs, intestines, and blood are responsible for 30 percent of the deaths within the poorest regions of the world, and tragedy during childbirth accounts for another 10 percent.

Nevertheless, being poor on planet Earth does not carry the death sentence that it once did. During the last twenty-five years, access to clean water has increased by 30 percent and access to improved sanitation has doubled in the poorest nations of the world. Just in the last thirty years, rates of immunization have doubled and access to prenatal care has increased by more than 30 percent in the same regions. The result is that the crude death rate in poor nations is now less than half of what it was when I was born in 1969, and as I mentioned above, it is now equal to what it is in wealthier nations. As with death during childbirth, we still have a ways to go, but we are walking the right path.

How, exactly, during the last fifty years did so much more of the world come to better fight disease and survive childbirth? They mostly did it the same way that my grandparents did it a hundred years ago: by packing up and moving into town.

4

Where We Are

At Rome you hanker for your country home;
Once in the country, there's no place like Rome.
—Horace, *Satire VII* (c. 35 B.C.)

How many generations must you trace back in order to encounter a farmer? In my case, I need look only to my mother's parents, both of whom were reared upon the bucolic pastures of southern Minnesota and northern Iowa. My grandfather, after returning from World War I, wandered the country at loose ends, finally stopping in Wright County, Iowa, just long enough to barter with my great-grandfather. He swapped a fixed term of his own agronomic labor for an entire lifetime of my grandmother's domestic labor, which was a standard midwestern contract of the time. This transaction would bear fruit in various forms, including the ten-plus offspring that my grandmother produced over the course of her lifetime—one of whom grew up to become my mother. Then, in 1920, town life called and they answered: my grandparents permanently relocated their young family to my hometown hamlet of Austin, Minnesota, and my grandfather took up indoors work in the hog kill.

If you have to look back more than eight generations to find a farmer, or a rancher, or a trapper, then you and your lineage are extremely rare and conspicuously urbane. In 1817, only 3 percent of the global population lived in any kind of a city. Now, just two hundred years later, about half of the people on planet Earth reside in cities—that is, within urban centers containing at least one hundred thousand people. Cities may hold half of the world's population, but they are also distributed very unevenly around the globe.

Understanding the geography of the total global population is fairly straightforward. The "highly developed" set of countries that include the European Union, North America, Japan, Israel, New Zealand, and Australia (sometimes known as the Organisation for Economic Co-operation and Development, or OECD), all put together, adds up to just over one billion people. Amazingly, there are now two countries in the world that can boast more than one billion citizens: China and India each contains about one and a half billion people. The portion of Africa that sits south of the Sahara desert is also home to about one billion people. Similarly, the miscellaneous countries of East Asia (not including China) add up to about a billion people. The Arab countries of North Africa and the Middle East are home to half a billion people—as are Latin America and the rest of South Asia (not including India). The exact total, as of 2017, came to seven and a half billion people on planet Earth.

When I was born in 1969, the great majority of the people on Earth lived in rural areas. Today, the great majority of us live in urban centers. This means that during the last half century, about three billion people were added to the world's cities. A fair number of these cities qualify as "megacities" that contain more than ten million people. Cities of such huge size were unthinkable one hundred years ago; today, forty-seven of them dot the globe. The original megacities continue to grow: Tokyo now has more than thirty-five million people, and New York City has exceeded twenty.

Cities are the very definition of *more*—hot spots of humanity visible from space. At night, they pulse with artificial light, and from above they resemble nothing so much as the network of nerves in the brain. Gleaming dendrites of suburbs splay from the blazing nuclei of city centers, each one connected to its neighbor by an axon of glittering highway.

A high density of people translates into the possibility of concentrated human labor: Henry George was right, back in 1881, when he wrote that as new people arrive, "the hands they bring with them can in the natural order of things produce *more*." The history of cities dating back to ancient Alexandria reveals libraries, construction projects, transportation systems, bazaars, hospitals, bureaucracies, judiciaries, and myriad other enterprises that could never have been endeavored within low-density population settings. Cities always have been, and always will be, places where people come together in an attempt to divide and multiply their labor.

For my part, I was born and grew up in the same small town where my grandparents had landed in 1920 (pop. 10,000). When I became an adult, I moved to an urban area with a population of more than one million, seeking education and work—a very common story the world over. During the last half century, just over one billion people either moved into, or were born into, a city that already contained more than one million people. I was one of the multitude who migrated into the city, and we all did it for more or less the same reason: in hopes of finding opportunities that we did not see at home.

The very poorest of the poor, the one billion people who struggle to survive on less than one dollar of spending power per day, are the people who were left behind in rural settings. More than three-quarters of people with no access to electricity— no light for reading after dark, no refrigeration for milk, meat, or medication—live in rural areas. Every year, several million people move into urban slums because access to essential medical services and improved wages is still, as a rule, higher than the rural alternative. In addition, slums can—and sometimes do—improve: since 2000, access to sanitation and clean water increased such that the urban living environment of more than two hundred million people no longer fits the United Nation's definition of a "slum."

The world's cities will continue to grow; on every inhabited continent, people are migrating from rural areas into cities. Even in places such as Europe and North America, where more than 80 percent of the population already lives in cities, people are still migrating away from the country. And don't forget that total global population is still increasing as well; a pragmatic estimate is that we'll break the ten-billion mark by 2100. More people in more cities requires more *more* in general, but especially with respect to the food supply.

Which begs a question: After the whole world moves to the city, who is left to run the farms? The answer: Almost nobody. And the almost nobody who currently grows the food that is keeping you and me alive is what we will talk about next.

Part Two

FOOD

The world moves on with giant strides,
Last year abundance, famine this;
You see to what a monstrous girth
The folks' necessities have swell'd.
— Henrik Ibsen, *Brand* (1865)

5

Growing Grain

If you can look into the seeds of time,
And say which grain will grow and which will not,
Speak.

—William Shakespeare, *Macbeth* (1606)

There is a heart-shaped portion of land that stretches east to west from Chicago to Cheyenne and north to south from Fargo to Wichita, and by "heart-shaped," I mean that it is shaped like an actual human heart and not a paper cutout. At its very center, just where the pulmonary artery crosses the aorta, sits Sioux City, Iowa, which may be the most unassuming city in the United States of America. I grew up not too far from there, in the very heart of the Heartland.

Education is rudimentary and incomplete in the Heartland. For example, I did not know that I was supposed to be ashamed of being from the Midwest until I moved to California. During my first semester of graduate school, I shared a lab bench with an Ivy League boy who asked me, upon learning of my humble origins, "Oh God, what do you even *do* there?"

"We grow your food," I answered, while evaluating his muscle mass in terms of its ability to shovel snow.

I was mostly right. While the Heartland makes up only 15 percent of the land area of the United States, it is home to more than half of its farm fields. I grew up "in town," but I also knew darn well that our town existed only to slaughter the animals that had eaten the grain that had grown on the farms that stretched for miles and miles, all around us.

Where are you from? I'm not asking where were you born, but rather, where you grew up. What was the first thing you remember seeing out the car window? Desert? Ocean? Plains? Mountains? Trees? Buildings? More to the point: If your world ever falls apart, and you lose everything, where is the place that you will turn to, and then return to?

I am from Austin, Minnesota, as were my parents and my grandparents. Austin has a big factory and a few small restaurants, mostly diners and fast-food places. The northern city limit is marked by the cemetery where I'll end up someday, probably sooner rather than later because one thing that Austin doesn't have is a sizeable hospital.

Like a lot of families in my hometown, whenever one of us had a doctor's appointment, we would drive forty miles northeast to the clinic and then back again. As the youngest, I was never given the option of not coming along. Pressed into the back seat, I sulked beside whoever needed doctoring and passed many sullen miles across many years, staring out the window and watching the farm fields go by.

In November, the fields were barren, their black soil powdered with frost. After winter descended, everything went white, and I was unable to pick out the horizon from the colorless expanse that surrounded us. In April, all that snow turned to slush in one day, and meltwater streamed through the ditches on both sides. During May, I cracked my window to smell the thick air of planting as giant mechanical cultivators groomed the fields into rows. Plowing back and forth at six miles per hour, they looked like solitary metal beasts pacing their territory and marking the path behind them.

By the first of June, the corn was seeded and the soybeans were in the ground. Springtime rains never fail in Minnesota, and soon tiny green blades of corn bristled up in rows. When pea-green ovals popped up one week later, we knew that the beans had taken. Then summer subsumed everything, and the crops rose up—inch by inch, foot by foot—regreening the world around us.

Corn and soybeans are strange bedfellows, often found adjacent to each other, but different to their core. Soybeans are harder to *not* grow than to grow: they get their nutrients from the bacteria trapped in their roots, while corn plants require lavish applications of fertilizer. Soybean flowers are self-pollinating and so develop embryos without help, while every single seed on a corncob must be pollinated individually. Soybeans are harvested from a pod, four seedy knuckles inside a green leather glove. Corn embryos harden and dry by the hundreds, lining the cobs that hang from a stalk. Once corn and soybeans are fully mature, however, they meet the same end, both of them destined for the silo.

In October, the tractors reappeared, this time dragging behemoth combines that gathered all and left nothing in their wake. By Halloween, the entire county was a carpet of uneven stubble, and the odd dry husk whipping about was the only evidence of summer's bursting glory. By November, the first frost had come, and one of us was sick, and we were driving to the doctor, and the cycle had started all over again.

Nowadays, whenever I get the chance, I drive those same forty miles from Mower to Olmsted County and back again. I always take the same road, the one that we traveled so often as a family. I scan the landscape carefully as I drive, searching for the familiar signs of planting, growth, and harvest. I find comfort in the things that will never change, even as I spend my days cataloging the things that have.

Unlike much of small-town America, my hometown is not a ghost town: it has thrived in its own little way. It is a few hundred buildings surrounded by an ocean of soybeans, alfalfa, and corn. Until I was seventeen, that square mile of town and the farm fields at its edges was the whole universe to me. But it turns out that much of what I saw happening in the farm fields of Mower County—barren, blooming, and bursting by turn—was also genuinely representative of what was happening the world over.

Today, the farm fields of my home county produce over three times more food than they did when I was born in 1969. Likewise, the farm fields of planet Earth produce three times more food than they did in 1969. A planet that once produced one billion tons of cereal each year now produces three billion tons. More amazing still is that the amount of land under cultivation has not significantly increased in America over that same time period, nor has it much increased all over the world.

How did we come to be growing three times more food on

only 10 percent more land? The answer has to do with gigantic increases in *yield*—the amount of grain produced per footprint of soil.

A bushel is a unit of measurement that has been in use for more than a thousand years. One bushel of crop fills a basket that would hold just over eight gallons of fluid; it's a lot to carry, but not impossible. A bushel of grain weighs fifty to sixty pounds, which is a bit over the maximum that you are allowed to check in your suitcase when riding on an airplane. Fifty years ago, a plot of land the size of a basketball court was required to produce one bushel of corn; today, all that is needed is two parking spaces.

Both wheat and rice show the same stunning increase as corn, their average yields more than doubling over the last fifty years. Soybeans, barley, oats, rye, and millet all show dramatic increases as well. Not only that, but coffee, tobacco, and sugar-beet yields have increased by more than 50 percent—in fact, every single field crop that I researched for this book showed a substantial increase in yield over the last half century.

With very, very few exceptions, every farm field on planet Earth produces *at least* twice as much food today compared with when I was a kid, and it's a good thing too, as the planet contains twice as many people today compared with when I was a kid. This amazing agricultural feat was actually the result of three separate but related accomplishments: we feed plants better now than we did then, we protect them better, and we have improved the plants themselves.

Every plant on Earth—including a tomato bush, a beanstalk, or a tuft of wheat—needs water and nutrients to survive and grow. A plant can obtain these important resources from only one place: the soil beneath it, a complex mixture of deceased life and degraded rock that has festered below ground for hundreds, if not thousands, of years.

Different types of plants have different nutritional needs, and every handful of soil contains a unique combination of offerings. The variable landscape of what is available to plants, coupled with the differing needs of different types of plants, is what gives rise to different ecosystems; grassland *versus* rainforest *versus* wetland, for example. In contrast, a farm field is a completely artificial landscape; it is obliged to support a "monoculture," a single, designated plant type. The soil of a farm field is forced to be the perfect environment for monoculture growth. This is achieved by adding nutrients in the form of fertilizer and water by way of irrigation. During the last fifty years, engineers and crop scientists have helped farmers become much more efficient at supplying exactly the right amount of both.

World usage of fertilizer has tripled since 1969, and the global capacity for irrigation has almost doubled; we are feeding and watering our fields more lavishly than ever, and our crops are loving it. Unfortunately, these luxurious conditions have also excited the attention of certain agricultural undesirables.

Because farm fields are loaded with nutrients and water relative to the natural land that surrounds them, they are coveted as luxury real estate by every random weed in the county. In addition, the fields that were plowed for planting were already home to innumerable insects, fungi, and bacteria that are only too eager to eat key parts of our crop plants before we can. To control these pests, farmers use "pesticides," chemicals that are poisonous to the weeds, insects, and microorganisms that would otherwise compromise the monoculture—and are sometimes toxic to human beings as well.

Today, more than five million tons of pesticides are applied to cropland each year the world over—that's a little more than one pound for every person on Earth, and it represents at least a tripling in the manufacture of pesticides since 1969. Different crops require different poisons: the greenhouses of the tropics are fogged with tons of chlorothalonil to keep strawberries from molding; the rice paddies of China, Japan, and Korea are sprayed with thousands of tons of chloropyrifos to keep insects from breeding; millions of acres are soaked in atrazine to keep weeds from invading the vast croplands of the middle latitudes; and an unthinkable volume of fertilizer is slathered just about everywhere. Thanks to this, our farm fields are the ultimate safe spaces for the plants that we choose to cultivate and harvest. Our crops grow tall and strong, blithely oblivious to all the weeds that would steal their precious sun and immune to all the things that would eat them who are also not us.

But even more important, and harder to explain, is how the corn, soybeans, wheat, and rice that we grow today are not the corn, soybeans, wheat, and rice of fifty years ago: they are, in fact, better versions of themselves.

The prized portion of a crop plant is an exaggerated tissue—it is a fruit, seed, stem, or root that is loaded with sugars, oils, and proteins and bloated in size compared with its wild ancestor. Have you ever picked wild blueberries, dug for wild potatoes, or plucked wild grapes? They are significantly smaller, and often less sweet, than their grocery store counterparts. The same is true for the grassy ancestors of wheat, rice, corn, and the other grains upon which we rely; domestication of nutritious grain from wild grass species occurred over thousands of years of deliberate selection performed by our great-great-great-and-more-grandparents.

The ancient practice of plant breeding accelerated dramatically during the early nineteen hundreds, after agriculturalists learned to apply the principles of crossbreeding that the famous monk Gregor Mendel had observed by growing yellow and green, round and shriveled peas in his monastery's garden. Within the research fields and greenhouses of the 1920s and 1930s, scientists purposefully mated individual crop plants and kept careful record of the stature and resilience of the offspring, relative to its parents. They created hybrids by cross-mating these cross-mated progeny and obtained unusual strains with further enhanced vigor; they triggered mutations in order to increase variation. For more than five decades, scientists genetically modified all of the major grains cultivated on Earth, as well as most of the fruits and vegetables, using only a basic picture of plant genetics as their guide. The effect was amazing: grain yields tripled between 1900 and 1990, but more breakthroughs were still to come.

DNA—deoxyribonucleic acid—is a chemical that is present in every living cell. The molecule itself is shaped like a twisted chain and made of individual links. Only a handful of different links exists in nature, and yet every individual—each mushroom, human, palm tree—contains a completely unique sequence of a couple of dozen links. The key to this uniqueness is the total length of each chain—fungus DNA is a chain with millions of links, human DNA contains billions of links, and plant DNA can approach a trillion links.

Morse code is made of just two types of signals that sound like long and short beeps to the human ear. Written out, a chain made of Morse code "dot" and "dash" links can communicate everything from a simple SOS distress call to all five acts of *Hamlet*. In a similar way, biological DNA codes not for words but for proteins—it is one long recipe book, if you will. We refer to the individual recipes as "genes": a sub-chain of links that describes how to make a useful protein. Every gene is a recipe for a protein, and every protein is capable of doing a job—some of them very important jobs.

During sexual reproduction, the DNA chain from two parents is patched together in order to form the DNA chain of an offspring. From this, a new individual results, one with a collection of genes that is not identical to that of either parent. The offspring keeps some protein recipes from each parent and loses others. The crop scientists of the early twentieth century crossbred plant-parents that possessed desirable qualities, hoping to concentrate those qualities in the offspring, and it worked wonderfully, even if they couldn't identify the exact genes in the DNA chain that they had modified.

One of the biggest breakthroughs of the twentieth century was a new method that allowed geneticists to map out an entire chain of DNA, link by link. They called this new field "molecular genetics," and it allowed scientists to pinpoint individual genes within long DNA chains. Then, in the 1980s, they perfected a new method to alter DNA—a method that didn't require the hassle of breeding parents and waiting for a new generation to mature. These new "recombinant" technologies allowed geneticists to directly edit the sequence of links in DNA: they could delete, copy, and paste genes, all within a living plant. They could even take genes from nonplant DNA and insert them into plant DNA, infusing a baby plant with recipes for proteins that it would never have figured out how to make otherwise.

The group of crops that resulted from these recombinant methods is known as "transgenics"—or, more commonly, "genetically modified organisms" (GMOs). GMO grains are equally as nutritious as their nontransgenic parents, but they also contain recipes for proteins that make them less water demanding and more resistant to pests than their predecessors. These plants are simply better versions of their parents, who were in turn better versions of their parents; the only difference is that this newest generation was engineered directly from its own DNA.

The fact that GMO plants have laboratory-altered DNA does not make them dangerous for human consumption; the National Academy of Sciences has formally reviewed the safety of GMO crops twice now and found that they pose no unique hazards to human health. The problem with GMOs is that the seeds are sold exclusively by the handful of companies that played key roles in their development. If you are a farmer who wants to grow these wunderkind varieties of soybean or corn in order to keep up with your neighbors, you will inevitably find yourself doing business with Monsanto and DuPont, knee-deep in a near-monopoly.

Today, just thirty years after their introduction, a full 10 percent of the world's croplands are planted with GMO crops. Almost all of the soybeans, corn, cotton, and canola growing in the United States are now transgenic varieties: all of them were invented after I was born. Much of the exploding green that I observed from my car window in Minnesota owed itself to the adoption of GMO soybeans and GMO corn within fields that had previously been planted with conventional varieties. After the introduction of these transgenic crops during the 1990s, the global yields of corn and soybeans increased by at least *another* 30 percent and are now more than quadruple what they were at the beginning of the twentieth century.

Every year, more and more GMO crops are being adopted around the world. Is it prudent that the literal seeds of global food production should rest in the hands of fewer than five American companies, to sell or withhold as they see fit?

Another problem with GMOs is the cat-and-mouse game involving pesticides and GMO crops. The compound glyphosate is the number one pesticide used in the United States. Glyphosate works by acting as a red light to the building of all new plant tissue. As such, glyphosate can be strategically applied to places where one wishes to stop all plant growth, such as the soil between crop rows in a farm field. Glyphosate was first sold under the trade name Roundup in 1974, and since then, almost two billion tons of glyphosate have been applied to U.S. farm fields.

In 1996, the same company that sells Roundup—Monsanto—began to market the seeds for "Roundup Ready" soybean, corn, and cotton plants. Into these new transgenic crop varieties had been inserted a gene that can green-light plant growth, thus overriding glyphosate's red light. Planting Roundup Ready GMO varieties made cultivation simpler: Roundup herbicide could be applied liberally to the entire field, not just carefully in between rows, and the weeds died while the crops thrived. Total use of glyphosate has skyrocketed since the adoption of Roundup Ready varieties, increasing glyphosate use by more than a factor of fifteen over the last twenty years. Our farm fields are now more drenched in pesticides than at any point in history.

But the story doesn't end there. As the wind blows across transgenic farm fields, pollen from Roundup Ready GMOs is carried into adjacent ecosystems, where it participates in fertilization. In 1998, there was just one recognized species of glyphosate-resistant weed; today there are more than fifteen. Pesticides, like all antibiotic substances, become less potent as their targets evolve tolerance to the toxin that they employ, and so emerges the age-old feedback: the more we rely upon them, the less effective they become.

There's another issue of even greater concern: in 2015, the International Agency for Research on Cancer determined that glyphosate (Roundup) is a "probable human carcinogen" leading to an increased incidence of non-Hodgkin's lymphoma. Because these risks are dose related, they almost exclusively affect the people who handle the largest amounts of Roundup: our nation's farmers.

Somehow, whenever I talk about agriculture, all roads lead to Iowa. It's not just that I grew up in a small town where most of the roads did, in fact, lead to Iowa. It's also because the state of Iowa has always done the heavy lifting for agricultural America. There were years during my childhood in the 1970s when Iowa came close to producing one-quarter of the country's number one crop. The state of Iowa always has been, and probably always will be, the most productive farm field in the world.

When I was in high school, more than 80 percent of Iowa was farmland. And today, thirty years later, more than 80 percent of Iowa is farmland. Iowa has never *not* been more than 80 percent farmland.

But back in the 1970s, Iowa's farmland was divvied up into almost twice as many individual farms as there are now, and during the decades that followed, the median farm got smaller, while the biggest farms became huge. In 1969, the median farm in Iowa was about two hundred acres in size. Fifty years later, the state's median farm size has shrunk by half, while the number of farms comprising two thousand acres or more (picture four Central Parks side by side) has increased by a factor of ten. The largest farms in Iowa are twice as big as the island of Manhattan and harvest more than a million bushels of corn every single year.

At present, there are just under ninety thousand "agricultural principal operators" (that is, farmers) in all of Iowa. That amounts to 3 percent of the state population responsible for nearly 10 percent of the state's economy. Heavy lifting indeed.

Speaking of money, if you'd like to pay more for a food item that is exactly as nutritious as its conventional counterpart, the "USDA organic" label allows you to do just that. The requirements for USDA organic certification are focused on the long-term maintenance of soil fertility and so have much to do with cultivation practices and soil testing. They also promote mechanical methods of pest control over chemical ones but do allow for the approved use of a long list of synthetic pesticides. Certification as "organic" is designed to help "restore, maintain and enhance ecological harmony"—note that none of the certification requirements involve an evaluation of the food itself. Independent researchers did find that the level of residue from the most toxic pesticides in use today is much lower on organic than on conventional produce, however.

Thus, over time, an assortment of "organic" foods that our national agricultural agency admits to being better for the land we till has become available as a high-priced option within our grocery stores, a parallel boutique system of agriculture that is approachable by very few.

As usual, I was both right and wrong when I told my college lab partner that the Heartland is where we grow his food. The truth is, we do a lot of things with the grain that we grow in the Heartland, but we don't much feed it to people.

Take corn, for example. Those of us who grew up in the Heartland already know that a cornfield produces more than grain. It is the set and stage of our lives, it is the living furniture of our backyards. It's where you play hide-and-seek on Sunday afternoon while the grown-ups are still eating potato salad and it's not yet time for pie. It's where you pick up two garter snakes at the same time, one in each hand, to prove to your brother that you aren't afraid. Eventually, it's the place where you park the old Ford at midnight and boast to your friends that you can't wait to leave this little town, while you stare up at the cold stars, wondering why growing up has to also mean leaving home.

The United States has always been a garden of corn. In 1870, just after the end of the Civil War, when all was chaos, America still produced one billion bushels of corn. By 1890, production was up to two billion bushels each year. During the economic boom that followed World War II, corn harvests approached three billion bushels.

By the time I was a child in the 1970s, America was regularly producing five billion bushels of corn every year—more than all its other major grains combined. Today, annual corn production in the United States is at fifteen billion bushels—a stunning 300 percent increase over the past fifty years. And this was accomplished by converting only 50 percent more cropland to corn.

Over the last fifty years, the massive accumulation of American corn has spurred the development of increasingly bizarre industries dedicated to getting rid of it, such that cramming it down the human maw has been demoted to an ancillary method. We eat processed corn not only as cornmeal but also in forms as varied as gum, acid, wax, and MSG (monosodium glutamate), as well as the more recognizable starches, sugars, and oil. Nevertheless, human consumption of corn uses up only 10 percent of the U.S. annual harvest. Where does the rest of it go?

Half of the remainder (that is, 45 percent of the corn that will be planted, fertilized, and harvested this year) will never be eaten by any living creature. Out of the other half, more than one billion bushels—enough to feed one hundred million people for an entire year—will be converted directly into manure.

6

Raising Meat

*Man and the animals are merely a passage and chan-
nel for food, a tomb for other animals, a haven for the
dead, giving life by the death of others.*
—Leonardo da Vinci (c. 1508)

Once upon a time, in Ukraine, there lived a cow. To be more
accurate, it was a steer—a male bovine that had been castrated
just after weaning. The steer was young and strong and beautiful
and had many admirers. My friend was one of them.

My friend is a journalist and she was traveling in eastern
Europe, documenting the important ceremonies of the local wom-
en's lives. Her tour guide for that day was a man preparing for his
sister's upcoming wedding. He drove my friend through a pictur-
esque green valley and stopped the car next to a small holding pen
that had been carefully constructed and quaintly decorated. In the
middle of the enclosure stood a lone steer, chewing contentedly.

"This is the cow that I will slaughter for the wedding,"

the man said proudly, beaming toward the healthy animal. He explained with great effort how everyone in the village would be invited to the celebration. Each guest would prepare her very best dish for the party, but the animal standing before them was destined to become the crowning glory of the feast.

My friend looked at the docile animal and thought of its death. The steer looked back at her for a moment, then lowered its head, deeply inhaled the earth beneath its feet, and took another bite of grass.

There are nine hogs for every person in Iowa. I hesitate to share that statistic because it might inspire visions of a household populated by Mom, Dad, two kids, and thirty-six cavorting pigs. This image, delightful as it may be, is not an accurate portrait of life in Iowa. The unromantic truth is that the vast majority of Iowans do not know even one of their twenty-two million oinking neighbors personally.

These days it is exceptionally rare for an American to meet her meat, even though on average she eats ten food items derived from meat every single day. It is also not unusual for this meat to have come from the bodies of at least ten different animals. About one million animals are slaughtered for food every hour in the United States. This slaughter mostly takes place within airport-sized buildings, and huge swaths of the country specialize in different types of killing. Thirty million cattle are slaughtered each year in the Great Plains of Nebraska, Colorado, and Kansas. A whopping nine billion chickens are slaughtered each year across the Feather Belt that stretches from Arkansas to Georgia. One hundred and twenty million pigs are slaughtered each year in the upper midwest states that surround Iowa.

I hail from the virtual epicenter of the pork industry. My hometown may not be the cradle of porcine civilization, but it has a pretty good shot at becoming its grave: fully 6 percent of the hogs slaughtered in the United States each year breathe their last within the city limits of my tiny hometown. Between Fourth and Eighth Streets Northeast, thirteen hundred people kill nineteen thousand hogs every single day.

I've never physically been in the slaughterhouse; it's not like you can just knock on the door and visit. Our elementary school used to take a field trip through it—my oldest brother remembers going—but they stopped this practice by the time I got to third grade. Nonetheless, I know some people who work in the factory, and quite a few people who used to, so I can describe my hometown hog kill in reasonable detail at dinner parties. The other guests' reaction to these stories is usually something like pity—for me or for the pigs, I can't say—but it is not envy, I can tell you that much. Slaughter, and the housing of it, is at least sad if not downright ignoble, or so says the majority of people with whom I've eaten dinner over the last twenty years.

Puzzled and a little defensive by their reaction, I use my insider knowledge to try to convince them that my hometown hog kill is actually a good and decent place—or at least a benign place full of normal people. I explain how Temple Grandin herself had helped to design the facility such that pigs might be crossed over the river Jordan at a rate of one every five seconds with minimal trauma to everyone involved. While shuffling through the long, snaking lines (not unlike what we do at the airport), the hogs are totally unaware of their impending promotion to Glory, I assured everyone. Hell, they probably have to walk it off in Hog Heaven for a good long time before they even realize what just happened.

As for the people working in the slaughterhouse . . . well, killing hogs pays decently, I explained—much better than waitressing at the truck stop, for example. Full benefits, too, and there's even a clinic inside the factory, where workers get free medical treatment for both illness and chronic conditions. It's true that I've never met anyone who worked in the hog kill who also claimed to like their job—but then again, we were not exposed to the idea that one can both enjoy doing something and get paid for doing it until quite late in life, and we remain, more often than not, skeptical of the concept.

I try my best at those dinner parties, but I never make any headway in convincing people that Quality Pork Processors of Austin, Minnesota, is both a civilized and necessary concern. Upton Sinclair had gotten to them long before I did, and none of my descriptions of the present-day industrial process could displace his portrait of *The Jungle* from 1906. The really absurd thing about these conversations is that we were usually eating meat while having them. More than once, I told these stories *in between bites of pork*. Sometimes it was prosciutto, a lovely Italian ham that is not manufactured in my hometown. Never was it Spam, an unlovely ham-esque semisolid that is manufactured in my hometown. Spam was *invented* in my hometown.

Yes, most of the seven million pigs that visit Austin each year end up leaving as Spam, which is then eaten in eighty different countries at a global rate of one can every seventy-eight milliseconds. I never get around to sharing that statistic at dinner parties, yielding as I must to the tyranny of "let's let someone else talk now" culture, but I still insist, whenever I can, that we need to talk about meat more. We need to talk about meat because we are eating so incredibly much of it.

Since 2011, global production of meat has exceeded three hundred million tons per year—this is three times the amount that was produced in 1969. Ninety-seven percent of this meat is the product of just three pitiable animals: cattle, chickens, and pigs—the same three that made up almost 90 percent of our meat fifty years ago. The burden of progress, however, has not been borne evenly: to get twice as much beef, we slaughter only half again as many cattle as we did in 1969; to get four times more pork, we butcher only three times more pigs; to get ten times more chicken meat, we slaughter only six times more chickens. Add to this the fact that the world's hens now lay more than a trillion eggs each year—that's quadruple the production of 1969— and one begins to suspect that the twenty-first century will go down as a very dark period in the history of chickendom.

The trend holds true for other animal products too. Milk is a particularly shocking example: the United States produces twice as much milk as it did in 1969, but we get it from three million fewer cows. What's going on?

The answer has to do with *yield,* and it's not dissimilar to the explanation for how we are able to grow three times more grain on only 10 percent more land. We feed animals better now than we did then, we protect them better, and we have also improved the animals themselves.

Veterinary medicine, advancing through research, has brought new cures for animal disease and a better understanding of animal nutrition to the task of producing meat for market. More amazing still are the changes in animal physiology wrought by seven decades of controlled breeding programs within state-funded agricultural research and experiment stations. Thanks to the stunning work of a few thousand dedicated scientists, every single cow, pig, and chicken slaughtered the world over is, on average, between 20 and 40 percent bigger than it was in 1969. More meat from less biology is the unlikely prize conferred by the convergence of a bewildering assortment of seemingly con-tradictory characteristics, including quick maturation, high fer-tility, and low metabolism.

I suppose that the meaning of childhood changed for everyone during the twentieth century, but for nobody more than calves. During the 1950s, a calf commonly passed the one-hundred-pound mark during its third month of life; today, they routinely break two hundred pounds in just fifty days. Today's milk cow produces more than six gallons of milk each day, which is both double the amount of fifty years ago and a statistic that can only be fully appreciated by someone who has spent time breastfeeding another being.

Yes, the American family has changed, but more extensively than we realized. Mother pigs in particular are putting in long days: in 1942, the average number of piglets weaned from a litter was five; today, it's ten—this is compounded by the fact that sows now have babies not once but twice a year. And as for the humble roasting chicken, he is a fundamentally dissimilar organism to what he was sixty years ago: the amount of feed required to keep a two-pound chicken alive in 1957 is the same amount that keeps a ten-pound chicken alive today.

How many goodly creatures are there here! O brave new world over which we have dominion.

It took me some time to convince my journalist friend that when her Ukrainian host stopped his car beside that paddock, he wasn't doing it in order to show her an animal. He was showing her three years of hard work.

There's no such thing as a low-maintenance farm. A healthy steer, ready for slaughter, first gestated inside a heifer that required extra feed and tending for the 280 days of its pregnancy. After the calf was born, it required gelding, and after that came eighteen months of hauling hay, shoveling manure, moving pasture, fixing fences, watering, deworming, and—at the end—fattening. When my friend's guide chose his sister's wedding feast as the destination for all that meat, he was also gifting her the years of labor that he had performed in order to bring that steer up to the knife.

The production of meat requires a tremendous investment of resources; it is a process of concentrating an almost unimaginably large amount of raw materials into a relatively minuscule product. A full 30 percent of the fresh water used by humans on planet Earth is spent in the production, maintenance, and slaughter of meat animals. The twenty-five billion cows, pigs, and chickens that live in captivity while waiting to be slaughtered are given medicine in vast amounts. As of 1990, two-thirds of the antibiotics used in the United States are fed to meat animals, ostensibly to promote growth and decrease mortality, though studies have shown that they don't work for either.

Most of these medicines pass through the animals unabsorbed, mixing with their urine and then into the runoff that leaves the farm. The antibiotics seep down into the groundwater and provide training exercises for microorganisms; there in the deep, bacteria study how to resist our pharmaceutical defenses using the ancient microbial art of intergenerational trial and error.

The main thing that you must pump into an animal in order to produce meat, however, is grain—a veritable mountain of grain. More than sixty billion bushels of grain—mostly corn, soybeans, and wheat—are fed to meat animals every single year. When an animal eats grain, its body uses the grain for many things, but mostly not to build flesh.

Whenever a farm animal moves, breathes, vocalizes, excretes—every muscle that contracts, every neuron that fires—it uses energy gained from burning the food that it ate. It is possible to minimize loss of food energy by restricting movement, which has given rise to battery cages and gestation crates—spaces so confining that the animal within may not even turn its head from side to side. Even inside these contraptions, six pounds of grain fed to an animal results in only one pound of meat that may be harvested, as a rule.

In the world today, a billion tons of grain are eaten by humans while at the same time, a different billion tons of grain are offered to animals as feed. From all this feed, we get about one hundred million tons of meat and three hundred million tons of feces.

Can I confess something? I am tired of debating the morality of animal slaughter, and I don't tire easily. While the act of butchering is not more wrong today than it was fifty years ago, it is certainly more vigorous. As a species, human beings will slaughter six times more animals for their meat this year than they did back in 1969, and a full 10 percent of this killing will take place in the United States. Regardless of whether you subscribe to Genesis ("have dominion over . . . all the earth, and over every creeping thing"), or the Gospels ("all beasts upon the earth draw us to the kingdom in Heaven"), or none of the above, our planet is at a crossroads where your personal actions as to meat matter tremendously. And while an action may be grounded in morality, it doesn't have to be.

If every American cut their red meat and poultry intake by half—down from four to two pounds a week—it would free up

150 million tons of grain. This kind of decrease wouldn't constitute much of a sacrifice: two pounds of meat per week per person is generous compared with the consumption within many developed countries—it is still well more than the average consumption in Ukraine, for example. And the grain not used to make unnecessary meat for America would increase the world's food-grain supply by a respectable 15 percent.

If the thirty-six countries of the OECD (including North America, Europe, Israel, Australia, New Zealand, and Japan) together decreased their meat consumption by half, the world's food-grain supply would get bumped up by almost 40 percent. Or to put it another way: If the entire OECD adopted the habit of just one meatless day per week, an extra 120 million tons of grain would be available to feed the hungry this very year.

We currently share the earth with more than eight hundred million undernourished human beings. They are children, women, and men whose food intake falls "below the minimum level of dietary energy consumption needed to support the activities of life"; in other words, they are starving. There is absolutely no reason why anyone should have to live—or die—this way. Starvation is caused by our failure to share what we produce, not by the earth's ability to provide. Even with meat production siphoning off one-third of the world's edible grain supply, we still produce twenty-nine hundred calories of food for every one of the seven and a half billion people on Earth, every single day—that's more than enough to supply the per capita energy requirements for a healthy life, as set forth by the USDA.

The most conservative scientific predictions of global population growth put the world's total inhabitants near ten billion by the year 2100—that's an additional two and a half billion people at a minimum. All this means that you, and I, and the rest of the world have a little more than eight decades to figure out how to add another two quadrillion calories to the annual food supply—and to distribute it equitably—in order to avoid widespread famine.

Humanity has been in this exact jam before, back in 1950 when it became clear that the global population would eventually exceed four billion, and we wriggled out of it by vastly increasing yields. Now we find ourselves staring down the barrel of a long century having already played that ace. Science fiction notwithstanding, there is only so much corn that can be hung from a stalk and only so much meat that can be balanced on a backbone—we are running up against the biological limits of what our food can be. Sooner or later we will have to reconsider the fact that every year, we actively waste 90 percent of the grain we feed to animals, in exchange for a little meat and a lot of manure.

So much about the future is uncertain, but the fact that we need to eat will never go away. The total number of people is increasing, and we need to figure out sooner rather than later how we might feed them all. At present, we are choosing ourselves over our own grandchildren three times a day when we ignore the problem, pick up our forks, and take another bite of meat.

7

Finding Fish

All is fish that cometh to net.
  —Sixteenth-century English proverb

Norway is the happiest country in the world, according to every major news outlet in 2017, which is something to remind yourself when you're sitting outside on a frozen bench watching the guy on the other side of the train tracks pick bits of herring out of his beard and eat them. It is a country of five million people—that's smaller than the population of metropolitan Atlanta—and when I'm not waiting in the cold for a late train, I admit that I'm pretty happy living here as well.

I don't know if Norway has always been the happiest country in the world, but the fact that my ancestors voluntarily immigrated to Minnesota during the late 1800s suggests perhaps not. Things have certainly changed since then, and most Norwegians point to the 1969 discovery of massive quantities of oil underlying the North Sea as being part of the cause—but there's also a long history of global commerce at play. Millions of years before Statoil began harvesting fossil fuel from the ocean floor, the waters that surrounded Norway roiled with a different kind of export gold.

The North Sea teems with fat philandering fish; it does now and it always has. Norwegians are also nothing short of amazing when it comes to finding those fish and hauling them home. Twelve thousand Norwegian men and women currently consider themselves full-time fisherpersons, and they bring in almost 3 percent of the global annual catch. That's an average of more than two hundred tons of fish by each Norwegian fisherperson every single year.

Modern Norwegian fishing boats are about as high-tech as they come; touring one is like being inside an oddly rocking spaceship. Today's fleet, equipped with state-of-the-art Global Positioning Systems and underwater sensing equipment, captures the same amount of fish as did the fleet of fifty years ago but uses only one-third as many people and a lot fewer boats while doing so. In its own way, Norway's fishing industry was even more amazing back in 1970, when forty-three thousand Norwegians were bringing in 4 percent of the world's fish without the benefit of modern machinery or satellite-assisted navigation.

About thirty years ago, however, the Norwegian fishing industry became a whole new kind of amazing; and today, just seven thousand Norwegians supply all of Europe with its most popular fish.

It would be difficult to overstate the importance of the fish species *Salmo salar* on Norwegian culture. Known as the Atlantic salmon in English and simply *laks* in Norwegian, *Salmo salar* is a weather-hardened traveler, as all good Vikings are. Hatched in fresh water, it feeds on insects and their larvae, an avid hunter from day one. As an adult, it roams the North Sea, tracking squid, eel, shrimp, and herring, eating and growing into a behemoth that can measure one meter in length and a hundred pounds in weight. Later in life, it channels both the strength of Thor and the passion of Freya to struggle upriver, against the current, and return to its freshwater origins in order to spawn a new generation of piscine explorers.

Salmon flesh is stained with the rosy pigment of its conquered prey. It has been devoured in Norway for millennia as both a staple and a delicacy—raw, cooked, smoked, pickled, or dug into the earth by handfuls and left to ferment, then eaten as a hallucinogen to supply pre-battle courage. Where salmon go, Norwegians follow. The old Norse sagas, one thousand years old, tell how Grettir the Outlaw miraculously navigated his ship through the Stykkishólmur archipelago while exploring western Iceland. Having survived the treacherous eddies and whirlpools that swirl between those thousand tiny isles, his ship entered the open waters of the Hvammsfjörður. Gliding across the calm, protected fjord, he was overcome. A yellow-green forest of birch trees tinted the sunlight that bathed his face. When those trees parted, he looked up at the steaming Skeggoxl volcano and saw glacial meltwater streaming down its flanks.

Grettir came to the grassy basin at the end of the fjord, where two rivers shared one mouth and emptied themselves into the salty sea, and noted a multitude of fat, jolly salmon striving upriver to spawn. He named the place Asgard, convinced that his gods had summoned him into their home. It was this sweet, sheltered valley—its icy streams writhing with salmon—that he believed to be the geographic location of Heaven.

Today, 99.99 percent of Norway's Atlantic salmon comes from places that are marginally less sublime than Asgard, but on the other hand, there is now a truly otherworldly number of them. If you required the citizens of Norway to consume their entire domestic annual production of salmon, you'd be forcing each woman, man, and child to ingest a pound and a half of salmon every day. This might actually be possible: Thor once scarfed down eight salmon in a single sitting while cross-dressed as Freya and trying to seduce Thrym (it's complicated).

Instead, more than 90 percent of the salmon produced in Norway is exported, mostly to the European Union and particularly to France. Even so, the per capita consumption of salmon in Norway is high—one hundred times higher than it is in the United States, for example. Salmon is everywhere in Norway: it's not exactly cheap, but it's the cheapest thing that you can buy in your otherwise horrendously expensive Norwegian grocery store. It's also the least imaginative thing that you can request from your Tuesday-night fishmonger. But it didn't use to be that way.

When I was a kid in the 1970s, the global annual production of Atlantic salmon was consistently about thirteen thousand tons per year. Today, global production is getting close to three million tons—which means that production has increased by more than 20,000 percent. When I was growing up, salmon was fancy-dancy cuisine compared with cod or pollack or almost any whitefish. My relatives didn't serve salmon very often, and when they did, we had to eat all of it—even the skin (yuck). Nowadays you can buy salmon at McDonald's (at least in Singapore). What happened?

The short answer is that Atlantic salmon no longer comes from the sea. Fifty years ago, the world's fishing boats went out each year and pulled thirteen thousand tons of salmon out of the ocean. Starting in 1990, that figure began to drop, until this year only about two thousand tons of salmon will be fished from out of the ocean (and as ever, 10 percent of the global catch will be made by Norwegians).

In the late 1960s, Norwegians began to question the time, fuel, and risk involved in fishing for *Salmo salar*. What if, they asked, instead of searching the North Sea and catching only a few, we simply collect the eggs and fence them into a cage, fatten the hatchlings into fish, reach in, and pull out a harvest whenever we feel like it? Similar questions were being asked all over the world for all kinds of seafood—not just fish but oysters, squid, crab, shrimp, and lobster too. Why not grow these creatures in a cage, or a pen, or a net, or a lagoon, and farm them in place instead of chasing them all over the ocean? And so the industry of "aquaculture" was born.

In 1969, while I was busy being a baby in Minnesota, brothers Ove and Sivert Grøntvedt put twenty thousand baby salmon into a net near the island of Hitra on the western coast of Norway, and the salmon did not die. There, trapped within the calm waters of the fjord, they thrived. The brothers harvested them and made a profit that very first year.

By 1990, there were easily one hundred times more *Salmo salar* living inside floating pens in Norwegian fjords than were being fished from the ocean during an entire year. Five years later, Norwegian aquaculture had doubled its production of Atlantic salmon, and by 2000, it doubled yet again. Just two decades after the industry took off, Norwegian aquaculture had broken one million tons of Atlantic salmon per year.

Opening a Pandora's box of salmon has not been without cost, however. There are distinct advantages to traditional fishing, namely: the fish breed themselves, feed themselves, and succeed or fail as adults based upon their own efforts. When you raise salmon in a pen, you must hatch them, feed them, bathe them, vaccinate them, medicate them, deworm and delouse them, and anesthetize them so that you can examine them, tag them, and move them from place to place without injuring them. You have to paint the cages with copper in order to keep the mussels from building up and strain tons of feces out of the water, because fish excrete five times more than humans do, pound for pound.

A typical Norwegian salmon farm consists of six to ten cylindrical cages, each one anchored to the seabed and buoyed at the surface to support a massive closed net; they are about fifty meters in diameter and a similar length in depth. Inside each of these pens, which look to be about the size of an Olympic swimming pool from above, there swim upward of one million fish.

The cage is home to the salmon during the second and third years of their short lives; they are placed inside upon completion of their juvenility, which is spent on land inside a freshwater tank. During the twenty-four-odd months that they live netted in the pen, they are doused with fourteen pounds of antibiotics, two pounds of delousing agents, and twenty pounds of fish anesthetic. Fifteen thousand tons of fish meal is offered to them, and they produce about five thousand tons of feces. The salmon are harvested from the net and processed when they reach about ten pounds in size; each pen produces between three and four thousand tons of fish every year. The western coast of Norway is dotted with thousands upon thousands of these floating pens.

Norse sailors have been navigating the icy Atlantic Ocean for one thousand years, sailing back and forth between Green-

land and Norway since at least A.D. 1000. In the mid-1980s, however, Norway weighed the cost to the environment against the
incredible convenience and fecundity of aquaculture, and the
aquaculture industry won.

Prior to 1990, global capture of wild seafood was actually
increasing: between 1969 and 1990, it nearly doubled, from sixty
million tons to just under one hundred million tons. Most of
this total was in the form of wild-caught fish from all depths of
the ocean: cod and pollack from the bottom waters; tuna, mack-
erel, and herring from farther up. Many nations were noting a
decline in wild-fish populations as a result, so the idea of in-place
aquaculture seemed a panacea: a new font of dietary protein that
would feed a swelling population while easing the pressure on
our overtaxed oceans.

Thus 1990 marked a major shift in fishing all around the
globe, not just in Norway. Between 1990 and today, global seafood production has doubled, but the amount of fish pulled
from the ocean has not changed. More than half of the fish eaten
worldwide is now produced through aquaculture. This forms an
important contribution to the spectacular increase in our ability to produce food during the last fifty years: it is the fishing
counterpart to the increase in crop yields and meat production
that we saw in chapters 5 and 6.

In 1969, the people of planet Earth ate about forty million
tons of seafood, 85 percent of it as fish. Today, the total amount of
seafood eaten each year is three times higher, but much more of
it is now consumed as shrimp, crab, oysters, and other delicacies
that were six times less common in the supermarket fifty years
ago. The majority of these foods come from aquaculture; massive scallop farms sprawl across the mudflats of the East China

Sea, and a countless number of prawns ripen within the walled estuaries of Indonesia.

This great shift in how we acquire fish has not only changed what we can eat from the ocean, it also threatens to change what can live in the ocean. The relatively short digestive tract within fish means they require food that is very high in protein, compared with animals that live on land. To supply this protein, aquaculture facilities use a feed made from smaller fish that have been cooked, pressed, dried, and ground into meal. All of these small fish come from the open ocean.

To get one pound of salmon, you need three pounds of fish meal. To get a pound of fish meal, you need to grind up five pounds of fish. Thus, each pound of cage-raised salmon "costs" fifteen pounds of fish from the ocean. At present, about one-third of the total catch fished out of the ocean is ground up into fish meal and then fed to fish that live in pens. Anchovies, herring, and sardines are the most fished fish in the world, and almost all of the catch is used as fish meal for aquaculture. These small fish are also foragers, which means that they survive on plankton— the tiniest plants and animals of the ocean. Foragers are near the base of the ocean's food web, where they serve as a staple food source for more charismatic species, including dolphins, sea lions, and humpback whales. More small fish diverted into aquaculture to feed us means less food left in the ocean to feed them.

The most optimistic aquaculture experts reckon that it is probably sustainable to pull about thirty million tons of forager fish out of the ocean each year without collapsing the food chain. Right now, we're at about 75 percent of that figure. The United Nations estimates that humanity will want to consume an additional twenty million tons of fish, over and above today's level, by the year 2030. If we produce these fish by aquaculture at today's rate of efficiency, we'll need to be pulling something like twenty-eight million tons of forager fish from the ocean each year, which is just under the United Nations' thirty-million-ton limit. But where do we go from there?

The story of aquaculture is simply that of meat production, only underwater. As such, it echoes the story of meat production on land: the diversion of massive resources through a small space confining millions of animals who live short lives that end in our bellies. And, as with meat, every bite of fish that we do not take could free up many bites of food for someone else.

The last fifty years have also taught us that a fish cage isn't the only thing that you can tie to the bottom of the ocean.

To be specific, you can tie a cutting of seaweed—green, red, or brown—and it will cheerfully grow into double, triple, or quadruple its original size, after which you may harvest as much as you wish and start the whole process over again. There are thousands of acres of such floating forests off the coast of China, and though they are devoid of birdsong, they sway in the gentle breeze of the current and dapple the light beneath their canopy.

Seaweed has been on dinner plates in Japan, China, and Korea for more than a thousand years, especially *Laminaria*—better known as kombu—a brown seaweed, and *Porphyra*—better known as nori—a red seaweed that dries to glossy black. Yet another facet of the boom in shrimp, crab, tuna, and salmon brought about by aquaculture is that nori is now recognized around the world as the shiny dark ribbon that holds sushi together.

The global increase in seaweed production due to the

advent of industrial-scale aquaculture during the last forty-odd years is even more spectacular than that for fish. In 1969, just under two million tons of seaweed were harvested and marketed from the world's oceans; today, that number is closer to twenty-five million tons. Sushi, its global popularity notwithstanding, constitutes only a very small part of the total consumption.

More than nine million tons of the annual seaweed haul—that's almost half of the world's production—is not eaten by people. Part of it is dried, ground, and used as fertilizer for crops on land, another portion is used in feed for animals, and the remainder is processed into chemical additives for a dizzying array of products, including cosmetics, moisturizers, shampoos, toothpastes, lubricants, inks, and bandages. The other half is eaten by people, but not in any form that they would recognize as seaweed.

Seaweed is used to produce "hydrocolloids": very large molecules that dissolve in water to make a viscous solution. There are three main types: alginate, agar, and carrageenan; all are low-calorie carbohydrates that can be used to thicken almost any solution. Today, more often than not, when we purchase ice cream, whipped cream, and salad dressing, they were made using recipes that rely upon extracts of soy and seaweed to approximate the properties we associate with milk, eggs, and cream.

Seaweed extracts never become rancid the way milk- and egg-derived thickeners do, so they have been widely adopted to both thicken and stabilize foods that previously relied on animal products prone to spoilage, such as frosting, cream fillings, jelly, and jam. Alginate, agar, and carrageenan, all made from seaweed, represent three from about twenty new chemicals that have quietly revolutionized the entire concept of food during the last fifty years, in a way that has done us no good at all.

Seaweed hydrocolloids are partially responsible for something that is everywhere now but used to be rare when I was a kid: a package of food that can sit for *years* and wait for someone to eat it—and when it is finally opened, it is exactly as palatable as it was on the day it was sealed shut. As a result, candy, cakes, pies, and doughnuts are rarely more than a few hundred meters away from us thanks to convenience stores and vending machines. But there's another chemical, invented in 1969, that has since swamped the way we eat and, especially, the way we drink. It contains no vitamins, or minerals, or nutrients—only calories. It's making our lives, and the lives of our children, almost unbearably sweet.

8

Making Sugar

*The development of high-fructose corn syrup will
have no significant impact on either world sweetener
consumption or on world trade in sugar.*
            —Journal of Agricultural Economics (1978)

Sometimes I miss my father so much that I cannot taste my food. I am better now than I was, but there are days when it hits me especially hard, when I cannot believe that he is actually gone. He died in 2016, at the age of ninety-two. My mother and my brothers and I had sat wreathed around his frail body for days in the hospital, until the very end, when I crawled into his bed and held him while the doctor removed the ventilator from his trachea. His breath grew ragged and finally stopped, and I saw the nurse brush away a tear as she turned off the EKG monitor and unplugged it for the last time.

     I used to worry that I would forget my father, and that my memory of him would fade, but I have found it not to be the case.

I don't even have to close my eyes to see his face or to hear his voice, but it goes deeper than that. His smell, his breathing, his bodily presence, were what my infant brain studied while he held me as a baby—which he did for hours on end, "for no reason," according to my mother—and these things were what I knew before I knew my own name. If I am fortunate to live so long as to be overtaken by old age such that I forget what I am called, I will still recall the man who loved me first and who loved me best.

"Coke with chow; wow!" is what my dad would say *every single time* he sat down to a glass of Coca-Cola, his voice perfectly tuned to just the right frequency of incredulous enthusiasm—the same one that he used to recite Rudyard Kipling poems. Like many families during the 1970s, we didn't drink pop on a daily basis, and when we did, it was part of a makeshift supper when my mother was away from home. Coke with chow (wow) typically involved scrambled eggs and was the first dinner that I learned to serve when my mother was busy attending evening classes in nursing school.

The only thing that my father knew to do when he became hungry was to sit at the table and wait to be served; I never witnessed him turn on the stove, not once in my whole life. With that attitude, you can't afford to be picky or critical, and my dad was neither. "Coke with chow; wow!" he would exclaim every time his seven-year-old daughter set a plate of runny eggs before him, and every time his eight-year-old daughter made him a stack of burnt pancakes, and every time his ten-year-old daughter dished out watery pasta and then covered it in canned tomatoes. Then he would dig in, and all talking would cease, until he looked up appreciatively to inquire about seconds.

"Coke with chow; wow!" was the tagline of an advertisement that ran only briefly in 1956 but will ring immortal through the ages if my family has anything to say about it. My dad must have first heard the phrase thirteen years before I was born, but I heard him repeat it back to me for more than forty years afterward. Now I repeat the phrase to my son every time we sit down with our Pepsis at Target Field and wait for the Twins to start warming up. We chat while watching for signs of life in the dugout, and I tell him stories about his grandfather who used to say, "Coke with chow; wow!"—especially about the year when he turned eighty and often held his infant grandson for hours at a time, for no reason.

Once upon a time in America, most jobs required heavy physical labor. American women, in turn, spent considerable time preparing high-carbohydrate foods for their families. Frying and boiling go quickly, but baking and canning take time. Pies, cookies, and cakes require preparation plus a good hour or so in the oven. The straining, stirring, sterilizing, and sealing required for jellies, jams, and preserves can easily eat up a whole morning, if not a couple of days. When women began to work outside the home, the hours available for meeting domestic demands dwindled, and time-consuming activities such as baking got cut under the sensible rationale that they had mostly yielded treats, after all. Accordingly, the record of white sugar purchased per household shows a plummeting trend between 1950 and 1975. Across that same time period, however, the total amount of sugar that the average American consumed each day increased.

Some of the reason for this has to do with the irony that women were serving more desserts than ever across this time period; they were just serving them outside of their homes. During the postwar economic boom of the 1950s, more than one million new waitressing jobs were created. America's business-men were traveling farther for work, and eating away from home became common, as did the concept of the business lunch. By 2005, Americans were getting one-third of their total calories from restaurants.

But the largest source of new sugar in American history first appeared with the advent of "convenience foods"—a term coined by General Foods in the 1950s to describe their new line of foods and beverages that were "easy to buy, store, open, prepare, and eat." These ready-to-eat meals and snacks have come to dominate the supermarket aisles, gas station shelves, and vending machine racks of America. As of 2010, half of the money that Americans spend on food goes to convenience foods.

Convenience foods are loaded with sugar. Packaged cakes, cookies, and candies are more or less based on sugar, but sugar is also added to the cheese and sauces that season ready-to-eat meals, as well as to the sausage, bacon, and ham that go into them. At present, three out of four of the food items that Americans purchase have had refined sugar added to them in order to make them more attractive to the consumer.

In the 1970s, the average American was ingesting almost one pound of sugar per week in the form of sweeteners added to convenience foods. During the decades that followed, the American workday got longer. As families relied more heavily on convenience foods to meet their needs, the average intake of added sugar spiked to an all-time high of nearly one and a half pounds per week in 2004.

The year 1956—of "Coke with chow; wow!"—seems very far away now. It is difficult to imagine a time when a person's reaction to seeing a can of Coca-Cola sounded anything like "Wow!" In the sixties, Coca-Cola ads became more existential when statements such as "Things go better with Coke" replaced references to actual things such as "chow." Coca-Cola progressed to become "the Real Thing" in 1969, and by 1982, Coke was simply "It." In 1993, Coke's slogan was unsparingly categorical: "Always Coca-Cola." And by then, for many American refrigerators, this was more or less the case.

As recently as 2007, the average American was consuming one can of Pepsi or Coca-Cola every forty-three hours, and though consumption has decreased since then, every American man, woman, and child still drinks upward of one liter of cola per week, on average.

In fact, most of the rapid increase in sugar consumption between 1962 and 2000 came not in the form of food but as beverages: sodas, sports drinks, fruit punch, and lemonade— Americans increased from an average intake of one can of sugary beverage every other day in 1977 to drinking one can every seventeen hours by 2000. Today, sugared beverages are the cheapest (and emptiest) calories that can be purchased and make up a full 10 percent of all the calories that Americans consume.

But why did sugar-sweetened beverages, which had already been on the market for almost a century, take off so sharply during the last fifty years? The answer might surprise you: it all goes back to a couple of years of bad weather.

Fruit and honey have sweetened the human diet for millennia, but during the Middle Ages, merchants learned to refine and crystallize dry sugar from grassy cane plants in the sunny tropics and, later, from earthy beets dug out of the gravelly soils of Europe. Today, what we recognize as table sugar—brown, beige, or white—is a purified compound called sucrose.

Table sugar forms when two chemicals join together like a butterfly, one wing made of glucose and the other wing made of fructose. This butterfly, purified from sugar cane or sugar beet plants, is what Americans brought home in five-pound bags during the 1950s and regarded as a staple for their larders, but today it is no longer the dominant sugar of the American diet.

In 1972, a horrible drought parched the vast farm fields of what is now Russia; its breadbasket region, Ukraine, was particularly hard-hit. Normal precipitation did not resume for years, and sugar-beet harvests suffered tremendously. To offset these shortages, the U.S.S.R. sought to import both grain and sugar from the outside world and entered the global marketplace for the first time in decades.

On the heels of the drought in Russia, Hurricane Carmen blazed through the tropics, seven thousand miles away. She pummeled her way across the Caribbean Sea and into the Gulf of Mexico, devastating the sugarcane fields of Puerto Rico and Louisiana. During the years that followed, low global supply met with high global demand, and Americans witnessed a horrifying spike in table sugar prices. Behind the scenes of this chaos, American exports of a seemingly unrelated commodity quietly doubled in value. That commodity was cornstarch.

After corn kernels are pulled from the cob, they can be replanted, eaten whole, or milled into their basic components: oil, protein, and starch. Cornstarch is a chemical with a very simple structure, hundreds of repeating molecules, all in a line, like a string of paper angels for a Christmas tree. In the 1960s, Japanese food scientists perfected a method to slice cornstarch into its individual units, and a second method to twist glucose molecules into fructose. Interestingly, the end product was a syrupy sugar, not a crystalline one—a syrup containing extra fructose relative to table sugar. They named this product high-fructose corn syrup, more simply known as HFCS.

Flustered by the 1974 sugar shortage, America turned full throttle to producing HFCS from corn, a crop that it had always grown in great abundance. It didn't take many years for America to become the world's largest producer of HFCS; by 1982, the United States was exporting more than a thousand tons per year. HFCS began to displace table sugar consumption, and today, one out of every three calories from sugar in the American diet is consumed as HFCS.

The truth is that HFCS is superior to table sugar for many purposes. Crystalline sugar has a nasty habit of hydrolyzing when mixed with acids, which is a fancy way to say that it tastes funny and turns brown. Not only is HFCS free from this trait, it's also hygroscopic, which means that it holds moisture, thus helping to maintain the appearance of freshness over time—perfect for packaged treats that must wait for months, or even years, to be selected from a gas station shelf or the inside of a vending machine.

Most important, HFCS is a ready-to-dissolve liquid, making it an ideal sweetener for beverage manufacture. As the benefits of HFCS took hold, its production skyrocketed, and the amount of American corn set aside each year for milling into starch increased dramatically. In 2001, Americans were manufacturing more than nine million tons of HFCS per year from cornstarch and consuming 95 percent of it as convenience foods and sweetened beverages.

The steep increase in HFCS consumption from basically nothing during the 1970s up to almost 10 percent of total calories by the year 2000 has closely coincided with Americans' sharp increase in weight over the same several decades, sparking a bitter debate among scientists about the culpability of HFCS in the obesity epidemic. There are good arguments on both sides: HFCS has effectively loaded the American diet with fructose, and high rates of fructose intake have been shown to disrupt fat and insulin metabolism in animals. On the other hand, as HFCS intake increased, physical activity did not increase, leading to a simple excess of calories.

It is not clear that HFCS is worse for your diet than table sugar, but it is clear that table sugar and HFCS are both worse for you than eating nothing. There is no downside, and no risk, to eliminating their consumption in favor of water—and some studies have shown great benefits. This is why nutrition intervention programs have targeted soda, and many consumers have embraced the campaign. Carbonated beverage sales in the United States were down 12 percent in 2012 from what they had been in 2007, and total sugar consumption has decreased as well. Sugar is faltering in America relative to twenty years ago, but today's figure is still easily twice as high as it was when I was a child during the 1970s.

At this point, you may be asking: What do manufacturers do with the excess ten million tons of HFCS that Americans have stopped drinking? The answer is simple: They export it to Mexico, where soda consumption has become the highest in the world. In Mexico, the average person gains 12 percent of their total calories from HFCS-laden snacks such as sugar-sweetened beverages and packaged convenience foods, and this hits children especially hard: more than 80 percent of Mexican adolescents consume more sugar than the recommended value.

As of yet, however, HFCS is not widely consumed outside of the United States, Mexico, and Canada; it is a local North American product. Indeed, HFCS is really just the newest by-product of the perennial surplus of American corn. Almost 20 percent of the world's cornstarch is produced within the United States; should global demand render it profitable, the United States could easily double or even triple HFCS production and export, essentially overnight.

But let's get back to table sugar, or plain old sucrose, because we are still eating an awful lot of that as well. In 1969, the entire human population consumed about sixty million tons of sucrose. Since then, global consumption has nearly tripled. This year, the United States will import enough refined table sugar to fill Yankee Stadium three times over. Then what? What happens to all this sugar—and the meat, and vegetables, and grain, and eggs and cheese that we put on our plates? Where does it end up?

Forty percent of it, at least, goes straight into the garbage.

9

Throwing It All Away

Behold it no more,
But throw it to the ground uncaring,
Let the common sewer swallow its identity,
So that the stars themselves no longer behold it.
                              —Thomas Middleton,
                              The Changeling (1622)

Pig's Eye, Minnesota (zip code 55102), is not as glamorous as it sounds. Every winter, there's a ninety-day stretch where it doesn't get above freezing outside, and there's usually a week or so when it gets down to forty below. If you lived in Pig's Eye twenty years ago, and the sewer line under your boulevard burst, my brother was the guy who came in a big truck, dug down into the muck, and fixed it.

My brother also claims that he got promoted for answering the phone. The guys at the sanitation department don't flinch at the idea of wading into frozen filth, but they find the idea of talking to a human being distasteful. There, in the basement of the public works building, the phone used to ring, and ring, and ring, and ring. When my brother couldn't stand it anymore, he'd pick up the receiver. "Hello?" he'd say. After a while, the people upstairs assumed he was in charge because he was the guy who answered the phone. He swears that this is how he got offered a desk job.

Pig's Eye was founded in 1840 and named after its most illustrious entrepreneur, Pierre "Pig's Eye" Parrant, a bootlegger who traded hooch out of a caved-in embankment on the Mississippi River. One year later, a French Canadian priest attempted to rename the settlement "Saint Paul" and officially embarked upon the first of many valiant efforts to make Pig's Eye seem like a place where you'd want to live even if you didn't have to.

When I was a college student in Minneapolis during the 1980s, Saint Paul was known as the place where you went to die. In those days, the hamlet of Pig's Eye was home to more than ten thousand people over the age of eighty. All these years later, Pig's Eye has become the place where you go to be born. Young families, priced out of the Minneapolis housing market, cross the river in search of cheaper pastures. When they meet my brother, they get very excited to learn that he has lived in Saint Paul for more than thirty years and ask him which neighborhoods have the best prospects for gentrification.

"Just don't send your kids to Such-and-Such Elementary," my brother advises.

Why? they implore him. Low test scores? Drugs? Crime?

"Naw," he tells them. "I've seen the sewers underneath that place. They can't take much more."

About 10 percent (by mass) of the food that you eat exits your body as solid waste. It has been thoroughly transformed, of course, and contains a large number of bacteria that did you a huge favor by helping you to digest what you ate. Living well requires a massive population of these bacteria in your gut, but you'll get very sick if even a small population gets near your mouth. The average adult produces about two pounds of feces and four gallons of urine each week, and all of it must be transported far away from the point of production and rendered innocuous more or less immediately.

The three hundred thousand inhabitants of Saint Paul, Minnesota, produce about thirty-six tons of feces and 150,000 gallons of urine each day. I'll spare you my trademark comparison-to-familiar-landmark visualizations of this human excrement—no, actually I won't. That's enough poo to fill ten concrete mixers to the brim and enough pee to fill one hundred additional concrete mixers flowing into the sewers of Saint Paul *every single day,* where it then becomes my brother's problem.

Now that I've got you pondering the physical dimensions of human waste, we might as well complete the exercise. The daily slurry of digestive discharge produced by the residents of Saint Paul amounts to just one-tenth of 1 percent of the total amount of human excrement produced each day in the United States. All of this filth enters the sewers of America, which uses a network of pipes and pumping stations to carry it to various treatment plants, where it is allowed to stew within massive basins. The liquid is pumped into a separate tank from the settled and floating solids, and then both fractions are subjected to a combination of chemicals and filters until they are deemed clean enough to be released into the bay, or river, or marsh, or whatever the adjacent environment might be; sometimes the solids are dried, transported, and then burned.

On the seventh of June in 1984, close to three inches of

rain fell on Saint Paul during a single afternoon thunderstorm. The sewers that carried both household waste and storm runoff quickly overfilled, until the excess was forced up through curbside grates and out onto the streets of Pig's Eye. Over the next ten years, Saint Paul spent more than $200 million separating the conduits that carry storm runoff from those that carry household waste, the first upgrade to the city's sewers since 1938. The project was completed in 1995, more than twenty years ago.

I have another brother who is a civil engineer, and he just won't shut up about this trend: major infrastructure built in the 1930s, expanded during the 1950s, minor upgrades during the 1980s and 1990s, and not much since. America's bridges, railroads, sewers—they're not in good shape, and we're starting to see them fail outright. Old cities such as Philadelphia have it the worst: three thousand miles of sewer lines, with construction started in 1883 and completed in 1966 and since then minor upgrades only. It's true that the total population of Philadelphia hasn't changed much since 1980, but on average, Americans eat 15 percent more food each day, which translates into 15 percent more . . . well, you get my drift.

As far as sewers go, however, even the worst ones in America are superior to the global average. Back in chapter 2, I first exposed you to the fact that global population has doubled since 1969. This also means that the amount of human waste produced on planet Earth has more than doubled, given that food consumption has increased greatly within many regions. Our ability to deal with this waste has not kept up, and today, a larger number of people than ever before live within conditions of inadequate sanitation.

At present, more than two billion people on planet Earth have no access to any system that would remove and sanitize the human excrement produced where they live. Similarly, about one billion people live without access to drinking water that has been fully purified of sewage. This forms an important part of what we'll see more clearly in chapter 10: that a large fraction of the world has yet to share in even the most rudimentary benefits conferred during our decades-long global pursuit of *more*.

Any analysis of rotting waste quickly broadens to include refuse far beyond human excrement. In addition to their massive output of human waste, the homes, schools, businesses, and hospitals of the United States, all taken together, generate about eighty million tons of organic waste each year in the form of discarded fruits and vegetables, other food scraps, and garden clippings. The total for all of the OECD countries is about 150 million tons per year, and the total for all the other people on Earth, combined, is close to four hundred million tons per year.

This means that America, which makes up 4 percent of the world's population, generates 15 percent of the world's organic waste. The OECD countries, all taken together, constitute just over 15 percent of the world's population but generate 30 percent of the world's organic waste. We will see this again when we talk about energy: much of the angst expressed about overpopulation is a red herring for the fact that a very minor part of the earth's population has done and is doing the majority of its damage.

There are many points between the farm and the fork where food is wasted. Vegetables are rejected for being too big or too small, grain spills from conveyor belts, milk goes bad on the truck, fruit rots on display, meat expires in the package, and uneaten dinner buffets are scraped into the trash. The more we eat, the more we waste: in 1970, each American wasted one-third pound of food each day, on average. Today, that figure is two-thirds of a pound. Twenty percent of what American families send to the landfill each day is, or recently was, perfectly edible food.

The business consultants that analyze the efficiency of American supermarkets reckon that one out of every seven truckloads of fresh food delivered gets thrown away. The trucks pull into the loading dock, the crates come out, their contents are unpacked, the shelves are stocked and later destocked into the garbage can, then taken to the dumpster—often by the same set of employees, who then turn around and begin unpacking another truck that just rolled in.

I am honestly unsure whether to feel more depressed or hopeful about it, but the magnitude of our global waste is in many ways equal to our need. The amount of total grain that is wasted is close to the annual food supply of grain available in India. The amount of fruits and vegetables that is wasted each year exceeds the annual food supply of fruits and vegetables for the entire continent of Africa. We live in an age when we can order a pair of tennis shoes from a warehouse on the other side of the planet and have them shipped to a single address in less than twenty-four hours; don't tell me that a global food redistribution is impossible.

The problem of garbage shows itself in the numbers—the megatons of food that fester and rot—but there's more to it than that. There is great tragedy in our waste. Every day, almost one billion people go hungry, while a different billion people intentionally foul enough food to feed them. We gamble our forests, fresh water, and fuel on food that we have no intention of eating, and we lose every single time. We annihilate, pointlessly, a countless number of plants and animals that spent the entirety of their short time on this planet in service to our appetites. And in the end, it's also about us.

That half-eaten meal in the garbage—why did we plow that field? Why did we plant those seeds, and water them, and fertilize the soil, and thwart the weeds? Why did we drive the harvester, run the thresher, fill the silo? Why did we deliver the calf from the cow? Herd it to the feedlot? Carry it in pieces from the conveyor belt? Why do we fix the refrigerators, and design the labels, and calculate the vitamin C content, and pave the roads, and change the carburetors so that we can drive, drive, drive meat, bread, fruit, sugar in boxes, bottles, and packages to the stores, schools, restaurants, and hospitals? Why do we walk the aisles, examine, select, buy, slice, mash, season, and serve? We spend our lives on these labors—we wake in the morning and leave our homes and we work, work, work, to keep the great global chain of procurement in place. Then we throw 40 percent of everything we just accomplished into the garbage.

We can never get those hours back. Our children grow up, our bodies wane, and death comes to claim some of those we love. All the while, we spend our days making things for the purpose of discarding them. When we cast food into the landfill, we are losing more than calories: we are throwing away one another's lives. It is the ultimate demonstration of how our relentless pursuit of *more* has landed us, empty and exhausted, squarely in the middle of less.

Let's assume, for a moment, that we have a choice. Now let's ask ourselves: Is this how we want to live?

Part Three

ENERGY

Come, my friend, own how boonless was
the boon; say where is any aid?
                    —Aeschylus (c. 480 B.C.)

10

Keeping the Lights On

I am the people—the mob—the crowd—the mass.
Do you know that all the great work of the world is
  done through me?

—Carl Sandburg (1916)

I have just one thing in my possession that belonged to my grandmother: the 1929 Singer sewing machine that she used every day for decades, to sew pinafores and pillowcases, quilts and choir robes, blouses and bedclothes, to let down hem after hem after hem. I never met her or my other grandmother—they both died before I was born—but when I was a girl, I would work the same treadle that she had worked, while I dreamed of her and wondered what she would have thought of me.

My mother bought an electric sewing machine in the 1950s—a Singer 216G. With the benefit of electricity, she taught me the things that my grandmother had taught her. She taught me how to imagine a dress that isn't in the store. How to size myself and map the inches onto newspaper; to extend the pieces for darts and pleats. To measure twice, cut once, pin carefully, then sew. To use the blind stitcher and the buttonholer and to sew with the presser bar raised when you have no other choice. To keep your fingers away from the needle and your hair away from the wheel. To take the tag from an old dress and hand sew it at the back of the neck when you're done, so nobody knows it wasn't store-bought.

It was my mother who first taught me how to transform a dream into a real thing that functions in the world. She learned this from her own mother, on the 1929 Singer sewing machine that is rusting in the corner of my living room.

I like to talk about boiling water in my classroom, because it's something familiar, and the best lessons always start from shared experience. "Suppose you want a cup of tea," I posit, "how many different ways are there to heat the water?" I make a list on the board as I call on each raised hand. An electric kettle, a gas stove, and a microwave are suggested right away, but I ask for still more. A driftwood campfire, offers somebody, and then someone else mentions a coal-fed barbecue. When I taught in Hawaii, students often speculated that there should be some way to usefully boil water by exposing it to hot lava; others knew that you can focus a sunbeam to start a fire and asked whether the heat could be used to warm something but not burn it.

"Many ways to skin the same cat," I end the lesson, pointing to our list, "many different strategies for obtaining the same result." I close with the moral of the story: "The requirements for each technique are different—electricity, gas, wood, coal, sun— but the result is the same: energy for boiling water."

We use energy every day, not just for boiling water but for many things: to warm and cool our homes, to light our way at night, and even to play a song on the radio. Today, mechanical motors provide the energy that we routinely supplied from our bodies just fifty years ago: golf carts have replaced caddies, can openers plug into the wall, leaf blowers push leaves instead of rake them, and a thousand other examples, large and small, fill our kitchens, garages, offices, and factories. My grandmother's sewing machine is an example of a muscle-powered tool that was once used to perform a daily task. It was first replaced by an electric version, like the one my mother and I used. By now it has been almost completely relocated out of the home and into the factory. As for myself, I have not sewn together a dress in years.

The total amount of energy that people currently use every day is three times greater than when I was a kid in the 1970s, despite the fact that the total number of people on the planet has only doubled. Much of this energy is consumed as electricity, and its usage is rising fast: the total amount of electricity that people use each day has more than quadrupled over the last fifty years.

Americans are the world's heaviest energy users, consuming a full 15 percent of the world's energy production and almost 20 percent of the world's electricity, despite making up only 4 percent of the world's population.

Those of us who were children in the 1970s in America remember being admonished to conserve household energy. On Saturday morning television, *Schoolhouse Rock!* cartoons chided us to "turn that extra light out," and President Jimmy Carter ordered us to turn down the thermostat and put a sweater on, just as he did at the White House. After Ronald Reagan moved in, energy efficiency—not conservation—became the name of the game. During the decades that followed, our labor-saving machines (particularly cars, as we'll see in chapter 11) were successfully engineered to produce more work from less fuel. Since then, we have aggressively offset what could have led to a global decrease in energy consumption by vastly increasing the level of energy use within almost every sphere of our daily lives.

Today, we watch baseball games inside air-conditioned stadiums, elevators are the default method of ascending or descending more than one flight of stairs, and even a book on your lap can draw an electrical current. The net effect is that now not only the tools of our labor but the general objects of our landscape consume energy. Our pursuit of convenience, diversion, repose, or whatever has driven us to live this way is yet another type of *more* that has caused great changes during the last half century.

Every semester I assign the same homework to my class. "Starting tomorrow morning," I instruct them, "please note down each time you use electricity." My students go home and gamely attempt to do so.

They usually compile at least ten entries before their breakfast ends: flipping on the light switch, using a hair dryer, plugging in the toaster and the coffee maker, and so on. But many of them have given up by dinner, overwhelmed trying to document the barrage of tasks that fill their days at school, at work, and on errands. There's plenty of items that they miss: the streetlights at which they stop and go, the water heater that warms their shower, and the array of battery-assisted devices including not just phones and laptops but everything from the clock on the wall to the speedometer on the dashboard.

When class meets again, we divide our lives into four parts: work, school, recreation, and family. We ask ourselves how each of the four parts are heated, cooled, lit, and maintained. I ask them a difficult question: How much of the working, learning, playing, and being together that you do every day could be done in the absence of electricity?

Electricity is a miracle invention: it allows us to light the darkness in order to extend our useful hours after the sun goes down, it sterilizes the tools that our nurses and doctors use within hospitals, it allows us to communicate with loved ones who are far away. Ever since I was born, I have consumed these luxuries and taken them for granted, and I have lived surrounded by people who are doing the same. One sad fact is that the last five decades of these innovations have failed to bring much benefit to a large fraction of the world. If you are reading this, you are already familiar with the portion of the world that uses too much energy, but you probably know less about the large part of the world that uses too little.

Thirty years ago, there were more than one billion people on planet Earth who had no access to electricity. Today, there are still more than one billion people on planet Earth who have no access to electricity. Because the global population has increased by 40 percent over that same time period, the total percentage of people living in poverty has decreased significantly. Nevertheless, the population that suffers from want is still very large; one out of ten people on planet Earth lives in abject poverty.

The global geography of need is also strikingly unaltered across the last fifty years. When you examine the global statistics of hunger, sanitation, disease, and poverty, the continent of Africa, with its long history of resource extraction and exploitation, stands out from the others in terms of both the severity of its troubles and their lack of remission.

The Sahara, which stretches from Mauritania to Egypt, is

the largest and hottest desert in the world. It is a formidable barrier that effectively divides the continent of Africa in half; south of the Sahara there exist forty-eight individual nations, each with its own government and laws. The region employs six different major languages and perhaps one thousand additional minor ones; each of these tongues has given rise to countless stories, poems, and songs. In terms of energy-based development, however, sub-Saharan Africa has been utterly left out in the cold.

The people of sub-Saharan Africa number one billion in total, which is just over 13 percent of the global population. However, more than half of the people on planet Earth who live without electricity live in sub-Saharan Africa. The region is also home to one-half of all people who live without clean water and one-third of all people who live without a sanitary sewer system. The number of people who die from the diseases associated with unsanitary conditions is high as a result: 50 percent of people who die each year from communicable diseases such as malaria, cholera, and dysentery die in sub-Saharan Africa. Behind these depressing statistics is the fact that while the population of sub-Saharan Africa has more than tripled in the last fifty years, the value assigned to the goods and services produced in the region is both very low and not increasing.

Every day, women and men all over the world work for six, eight, ten, or more hours in order to produce something of value. The monetary value that is assigned to what each person produces, however, is a complex result of supply and demand, fashion and prejudice, protection and risk, history and greed. Sewing thirty pairs of jeans, performing a tricky surgery, and teaching a child to read are three different labors that could each occupy an entire workday, but the three products—a box of clothes, a successful operation, an enlightened child—are valued dissimilarly by the global marketplace.

At present, the total product of the labor performed each year by the entire seven and a half billion people on Earth is valued at about 80 trillion dollars. That's up from about twenty trillion dollars in 1969, which is a quadrupling over the last fifty years. Labor performed in the United States and the EU, a population just under one billion people, is valued as half of the world's labor. In contrast, the labor of the one billion people living in sub-Saharan Africa is valued at under $2 trillion. This translates into 13 percent of the world's population producing just 2 percent of the world's value. Half a century ago, when I was born, the labor of sub-Saharan Africa was also valued at just 2 percent of the world's total.

A different kind of trouble has beset the citizens of India, yet another billion people. Since I was born in 1969, the total value of the goods and services produced each year in India has exploded: it has increased by more than 1,000 percent, while the population of India has only doubled. Despite this explosion in market value, the key indicators of poverty have changed very little in India. Thirty years ago, one out of every five people living without clean water and one out of every three people living without a sanitary sewer system lived in India; today, those figures are still the same. Other quality-of-life data tell a similar story: there are hundreds of millions of Indian citizens who have yet to receive any benefit from the global wealth that is pouring into India.

India and sub-Saharan Africa are similar, in that both regions consume very little of the world's energy. Together, they make up one-third of the world's population but consume less than 10 percent of the world's electricity. The vast majority of the world's people with no access to electricity live in India and sub-Saharan Africa. If you had to light a fire every time you needed to cook or clean, and there was no lamp to switch on after the sun went down, how would you go to school or study? It is perhaps not surprising that most adult women who never learned to read also live in India and sub-Saharan Africa.

The rich nations of the OECD sprawl from one side of the globe to the other, occupy almost half of its habitable land area, and are filled with people living well. The sum total of goods and services produced each year by the 15 percent of the world that lives in the OECD is now valued to be worth double that of everything produced in the entire rest of the world. While producing this product, and while living in general, this 15 percent also manages to consume about 40 percent of the world's fuel and almost half of all the electricity that is generated the world over.

This extreme imbalance in energy consumption inspires a simple sort of algebra: if all the fuel and electricity in use today were redistributed equally to each of the seven-plus billion people on planet Earth, each person's energy use could be equal to the average consumed by people living in Switzerland during the 1960s. I've seen pictures of Switzerland that were taken in the sixties, and you know what? It doesn't look so bad. People standing around in train stations wearing thick wool coats or sitting at small tables drinking from tiny coffee cups. What we saw in chapter 6 with food, we now see with energy: all of the want and suffering in the world—*all* of it—arises not from the earth's inability to produce but from our inability to share.

What was only a faint drumbeat as I began to research this book now rings in my head like a mantra: *Use Less and Share More*. As we'll see in chapter 13, there is no magical technology coming to save us from ourselves. Curbing consumption will be the ultimate trial of the twenty-first century. Using less and sharing more is the biggest challenge our generation will ever face.

It is a bewilderingly difficult proposition, and therefore unlikely. It is also the only surefire way that we can start to get ourselves out of this mess.

11

Moving Around

A man travels the world over in search of what
he needs and returns home to find it.
—George Moore (1916)

Sometimes, when I am sitting on the tarmac at Newark Liberty International Airport, strapped into seat 18G and waiting to take off for Minneapolis (which should have happened an hour ago), I think about all the things I could be doing if I could *just get out of this plane*. Then I think about the two hundred or so people directly adjacent to me who are probably thinking the same thing. When and if we finally take off, I concentrate on not thinking about how we are seven miles above the ground hurtling through the air inside a spare metal tube. Then I lower the window shade and try to sleep.

If, instead of flying, all two hundred of us escaped from the plane into two hundred separate cars and drove, individually, from New Jersey to Minnesota, we would have collectively burned 40 percent less fuel than we ended up using for that one plane by flying all together. If instead of using separate cars we had boarded a single passenger train, the total journey would have required only half as much fuel as was required for the gas-guzzling airplane that saved each one of us fourteen hours of travel time.

Of course, the most fuel-efficient option would be for all of us to have stayed home in the first place, and I remember when Skype first became available in 2003 and the business world announced the near-term obsolescence of face-to-face meetings. It didn't go down that way, however, and since then, we all fly more than ever. Today, Americans take almost two million more flights per year than they did in 2003—and the majority of these flights are for business.

Flying on an airplane has to be the most resource-intensive way you can spend your day, outside of being launched from Earth's surface in a spacecraft. The average American car gets about thirty miles to the gallon; the average airplane gets something like four hundred feet to the gallon. The average plane goes five to ten times faster than many modes of travel, however: cars and trains max out at about eighty miles per hour in most countries of the world. This has made the airplane into the world's conveyance of choice for long journeys, and indeed, the average distance traversed during a flight is eleven hundred miles, which is about equal to the distance between Newark and Minneapolis. Twelve such flights in a row is enough to plop you down on the opposite side of the globe, sick to death of salted peanuts and Sandra Bullock movies.

Routinely traveling very long distances has somehow become an ordinary thing, even though it was inconceivable to our grandparents. In 1970, the world's airlines carried just over three hundred million people each year, and almost 90 percent of these flights departed from, or landed in, OECD countries. Today, a global fleet of twenty-five thousand airplanes touches down thirty-five million times each year onto the runways of the world's forty thousand airports. In doing so, these planes move almost four billion passengers away from home and back again. Ten times the number of people will be long-hauled by airplane this year compared with 1970, so that we can work or play against a change of scenery, at least for a little while.

High-speed rail—that is, passenger trains that exceed speeds of 120 miles per hour—is the only industry able to compete with the airlines for long-distance travel, but consumers tend to choose flying when the extra time necessary for land travel exceeds thirty minutes. From Tokyo, where both planes and high-speed trains are readily available, five-hundred-mile trips to Osaka are overwhelmingly done by train, while one-thousand-mile journeys to Fukuoka are almost always flown by plane, even though the three cities are connected east to west on the Shin-kansen "bullet train" line. We travel so much that a single hour of not traveling is worth a great deal to us.

The statement above is not quite fair, not only because railway options do not exist everywhere but also because many of the railroads that do exist are in serious decline. Trains are being decommissioned en masse: the total number of locomotives in the world has been slashed by 17 percent in the last twenty years, even as population increased by almost 20 percent. In most countries (with the exception of China), the amount of train track available has not changed meaningfully during the last forty years, which means that as the world's cities have grown, our ability to move within and between them by rail has not. The United States has witnessed a conspicuous divestment in its railroads during the last two decades: it is one of the few countries within which the total amount of track has actually decreased.

Perhaps more important, the global workforce that maintains and operates our railroads has been gutted: in the last twenty years, one out of every four of the world's railroad jobs has been eliminated, mostly during the course of privatization. The United States, which has never enjoyed a vigorous passenger railroad system outside of the New York–DC corridor, has declined substantially: since 1991, one out of every seven railroad jobs has been eliminated. The United Kingdom's railroads have fared worse, in that one out of every three railroad jobs disappeared over that same period; and Spain has decimated its railroads, having eliminated four out of every five railroad jobs in the last thirty years. As these national railroads degraded, their use increased: since 1991, the number of passengers, and the distance they travel, increased by 20 percent in the United States, shot up by 70 percent in Spain, and doubled in the United Kingdom. This is not a recipe for consumer happiness.

Speaking of unhappiness brings us naturally to cars. If I have a single bias that I should reveal as the author of this book, it is that I hate cars. If I have a single hope for this book, it is that other people will write to tell me about how (and why) they also hate cars.

I hate all cars, and I also hate specific cars. I hate them because they hate me. I don't trust cars, and I never will. I'm also a lousy driver. Which is obviously the cars' fault too. I have a friend who understands this, and because of that, he is very dear to me. The only thing that he and I ever talk about is the "crapcars" that have persecuted us over the years, which goes something like this:

"I had an '85 Datsun that idled so hard it shook the muffler off. You had to either gun it or drive through downtown with no muffler. Very loud either way."

"I had a 1978 Pacer with a passenger-side door that would fall off if you turned right. Developed a complex route of left turns to get to work."

"I had a two-cylinder Volare with an accelerator that would pull down and stick to the floorboard while you were on the highway."

"I had a two-ton Chevy with no windshield and a bad latch so you had to tie down the hood with rope. It would run with the key in the 'off' position, so you had to disconnect the distributor to park. Makeshift electroconvulsive therapy."

"I had a 1999 Isuzu that I saw profiled on *Entertainment Tonight* as the most dangerous car in the world. Showed it mangling fifteen crash dummies in one go. I knew it was crappy, but I didn't know it doubled as a coffin."

"Had a 1970 Gremlin that tried to double as a crematorium. Rear taillights would fill with dirty water when it rained and short the engine out."

And so on.

Because I hate cars more than the Devil hates Jesus, it does not make me happy to tell you that the world currently contains close to one billion passenger vehicles. I personally hate cars because all my life, I have watched helplessly as they torture the people I love—I grew up surrounded by crapcars; then when adulthood struck, they came for me. There are other, and more rational, reasons to dislike cars, however, and foremost among them is the fact that they are outrageously dangerous.

If cars were not useful, they would be regarded as a great social ill: more people die from road traffic injuries each year, the world over, than from murder and suicide combined. We make formal and good-faith efforts to condemn, eradicate, or at least minimize both murder and suicide, but we vigorously duplicate and distribute the automobiles that are prone to spontaneously slaughtering us while we operate them.

America remains by far the most car-intensive culture on Earth. In America, six million new cars are purchased every year; in 2017, the car population increased 50 percent more than the American population. When you visit New York City, it may seem as if all humanity is riding the subway, but really only 5 percent of Americans use any kind of public transportation on a daily basis. The rest drive almost everywhere they go.

We drive and drive and drive in circles, some of us all day every day. Where are we going? Mostly, we go to work and back. Eighty-five percent of American adults drive to their workplace, and three-quarters of those who drive, drive alone. Americans have also begun to work longer days and to drive even farther to get there. The average U.S. employee works twenty-one hours more each year than she or he did just a decade ago. In 2005, 15 percent of commuters drove for forty-five minutes or more, each way, to their place of employment; today, that figure is closer to 20 percent. We also vacation and visit and sightsee, but most of what our cars do is take us away from the people we love, so that we can do the things that we have to do in order to buy more gas to put into our cars.

The total cumulative distance that Americans drive each

year is staggering: in 2015, it was equivalent to five hundred round trips from planet Earth to the (former) planet Pluto. Every day and on average, each American spends one hour in a car. Distance-wise, every woman, man, and child in America will circle the globe by car at least once during the next three years. We spend a good part of our lives in our cars—up until they maim or kill us, that is.

Let's pretend I've convinced you, that your eyes have been opened to the fact that cars are a murderous joy-sucking plague on the human species. Given where you live, could you give yours up, even if you wanted to? Could you secure the basics—food, education, medical care, a paycheck—on foot, by bicycle, by public transportation? For the majority of America's households, it wouldn't just be burdensome or impractical—it would be logistically impossible. The ground rules of American society necessitate cars, and those rules are less flexible than we'd like to believe. While pursuing the ever-elusive *more*, we inadvertently trapped ourselves into metal boxes. Now we spend our mornings and our evenings maneuvering among the other metal boxes, watching each other through the glass.

One hundred years ago, there were probably fewer than a thousand cars in the whole state of Minnesota. My dad grew up there during the 1920s and remembered the adults around him being skeptical of the idea that cars could ever replace horses as a means of daily transportation.

All that money for a vehicle, and impossible to know if you were getting a good one? They shook their heads, unable to imagine paying hundreds of dollars for a hunk of metal that you couldn't prod the haunches for muscle, or check the manure for worms, or even inquire as to the sire and dam. No, it would never catch on, they assured one another, plus where were you going to get all that petrol from? They'd have to sell it up and down the streets or bring it around in a milk wagon, and Main Street already managed to burn itself down once every five years without the benefit of gasoline running through the gutters.

A century later, there is one passenger car for every two people of driving age in the United States, and gasoline is indeed sold up and down the streets of America. The car as a working machine has changed mightily since the Ford Model T was introduced in 1908 as the first family automobile. The goal of the automobile industry has always been toward increasing "horsepower"—a measure of mechanical prowess created for the blatant purpose of converting families away from animal transport. The Model T boasted an engine with twenty horsepower: imagine a machine that could pull as hard as twenty Old Blazes, that never needed the water trough, and that didn't get barn-sour in its old age—not to mention the end of shoveling manure. *No more craphorse!* they exclaimed in their tragic innocence, and were finally converted.

As the speed and stamina of cars increased during the postwar industrial boom of the 1950s, the amount of gasoline required to fuel the automobile engine increased as well. By 1974, the average American passenger car could sustain speeds of fifty-five miles per hour on the highway (the speed limit of the time) and burned a whole bathtub's worth of gasoline every day that it did so. Almost all of America's imported oil came from the twelve member countries of the Organization of Petroleum Exporting Countries (OPEC), mostly located in the Middle East and North Africa.

When the members of OPEC proclaimed an embargo on

petroleum exports to the United States in 1973, gasoline prices quadrupled virtually overnight for American consumers, and the whole country was shaken into acknowledging its utter reliance on one imported commodity. As a response, America's auto engineers redesigned the passenger car toward alleviating this dependence, and by 1980, the average car engine was 50 percent more efficient and got about twenty miles per gallon, mostly because the cars themselves were, on average, 20 percent lighter than they had been just five years earlier (it takes less fuel to move less metal).

Combustion engines were greatly improved during the 1970s, and massive benefits emerged from the car industry's quixotic quest for perfect fuel efficiency. Airplanes were made 40 percent more fuel efficient between 1970 and 1980, and after oil prices descended to their previous levels, airlines expanded vigorously to welcome an ever-broadening swath of America on board for long-distance travel.

About this time, a momentous fork in the road appeared, illuminating a possible path toward the future. Americans could have remained one-car families and continued to shop locally and fly sparingly, and our total energy use would have decreased over the next several decades as engineering further enhanced the efficiency of our motor vehicles. As we'll see, this was the road not taken.

Today's automobiles average about 230 horsepower—more than ten times the poor old Model T and more than double the average of the 1980s. The average weight of American automobiles rose sharply as families put minivans and SUVs and pickup trucks into daily commuter use. By the year 2000, the highways were filled with vehicles every bit as heavy as the Lincoln Continentals and Ford Gran Torinos of the seventies, and at twenty-three miles per gallon, average fuel efficiency was actually lower than it had been twenty years earlier.

It didn't stop. Americans bought still more cars and kept driving them, and today they travel twice as far by car each year as they did in 1970. As a result, America's twenty-first-century dependence on foreign oil is higher than it has ever been, while its relations with the countries of OPEC are perhaps the worst we've known. The road not taken was also a Story of Enough, but when given the choice, we recommitted ourselves to the Story of More: America thoroughly offset any oil independence it could have gained from the miracle of renewed fuel efficiency during the 1970s and 1980s by doubling down, making more automobiles, and proceeding to use them harder.

For every year that I've been alive, Americans have added forty billion miles to the grand total number of miles they put on their cars while traversing the country's streets, roads, highways, and interstates. But Americans are not the only ones driving. Just since 1990, the people of China have added almost as much driving, to the tune of thirty billion miles per year. Outpacing everyone are the people of India, who have added a staggering three hundred billion miles behind the wheel per year, since 1990, to their annual total.

The total increase in driving between 1970 and today, for the above three countries—the United States, India, and China—necessitated a minimum of four hundred billion gallons of fuel: a volume so large that it would take twenty-four gushing Mississippi Rivers one hour to fill it.

The slippery black oil that we slosh into engine after engine after engine has its own story, sprawling and fanciful in its own right. But to tell it correctly, we have to back up, millions of chapters before page one.

12

The Plants We Burn

But I will give men as the price for fire an evil
thing in which they may all be glad of heart
while they embrace their own destruction.
—Hesiod (c. 700 B.C.)

Once upon a time, there was a broad and fathomless ocean. Beneath the waves swirled the currents that flushed salt water in magnificent gushes. Deep in the dense, cold darkness, the sea-floor waited with perfect patience to embrace the corpse of every living thing above.

This was called the Panthalassa Sea, and I am certain that the creatures swimming in it believed their watery world to be the entire Universe, extending forever in every direction. The creatures of the Panthalassa were different from the ones we recognize in our oceans: jawless fish writhed in great schools, their wormy bodies streaming behind them as they searched for places to attach. Innumerable cartilage forms swam, forever in motion, their gill slits pumping as they moved. Spiral-shelled squids jetted back and forth, contracting and squirting their way along. Sea lilies swayed, tethered to the bottom by their roots, their animal petals filter feeding on the small and weak that floated by.

The Panthalassa Sea was enormous—far bigger than our Pacific Ocean—but it did not extend forever. On the other side of the planet lay the Tethys Ocean, a thick, salty soup of green algae and the tiny animals that grazed upon it. Within this green slime, bubble-encased amoebas drifted from meal to meal and spongy coral sat anchored, waiting for dinner to pass through. Each tiny alga opened itself to the sun and was fertilized by sloughage from faraway mountains.

Between the Panthalassa and the Tethys there stood a vast continent, drained by rivers, pocked by volcanoes, and home to a different sort of life. Near the middle of this continent stood the Forest of Ruhr, which was altogether unlike the forests we know.

The trees of Ruhr stood thirty feet tall, crowned with fern fronds, spilling seeds from their base. Taller still stood tufted starbursts of needles atop hundred-foot trunks. None of these trees were made of wood: each one contained a hollow cavern of pith. Nor were they covered in bark: their rinds were marked by a paisley pattern of scars. Stranger still, no flowers could be found anywhere—not a one—nor the fruit that follows the bloom, nor the pollen that precedes it.

A mossy carpet lined the forest floor of Ruhr, and within it great millipedes chawed dead foliage and expectorated into the soil. Creatures strange and naked, turtles without shells, mucked their way through the fallen leaves, while dragonflies big as seagulls hovered above in the humid air. Like the sediment at the bottom of the oceans, the mud beneath this boggy forest quietly accumulated, accepting and incorporating the dead year after year, without judgment or prejudice.

The story above, for all its romance, is not a fairy tale. The Panthalassa Sea, the Tethys Ocean, and the Forest of Ruhr are all real places that existed between two hundred million and three hundred million years ago on planet Earth. There is not a drop of fiction in my descriptions, for ample evidence has been preserved in museums. There, you can wander through the history of this long-ago world, examining its stones and bones.

The Forest of Ruhr grew lush for millions of years; it was one of the earth's first tropical rain forests. Eventually, it was buried under the accumulating sands of time. Crushed and rotted, the sludge descended into the deep, where heat and pressure transformed what had once been leaves and stems into black, blocky coal. All the while, the entire continent was pushed and pulled until its heaving journey stranded it far from home, on the other side of the globe.

Similarly, the Tethys Ocean was folded down into the earth, and over tens of millions of years, the plants and animals that had died were cooked and pressed into a rich black broth that bubbled in seeps between sandy layers, until a thick crude oil was all that remained of the life that had been. The plants and animals of the Panthalassa Sea followed a similar fate, but their oily remains were abused further, until whole molecules cracked under the strain to produce bubbles. The bubbles slid along the rock layers and merged into great pockets of natural gas.

The plant and animal corpses of the Forest of Ruhr, the Tethys Ocean, and the Panthalassa Sea now constitute the coal deposits of Germany, the petroleum crude oil of Saudi Arabia, and the natural gas deposits of North Dakota, respectively. The largest coal reserve in western Europe was once a tropical forest; the crude oil that gushes from the world's most productive oil wells once lined a shallow ocean; the natural gas collected during today's fracking boom once anchored a deep abyssal sea.

A fossil is everything, or it is nothing, or it is something in between. An insect trapped in amber and a dinosaur's footprint— they both count as fossils, remnants from a time gone by. The bulk of life on land and in the ocean is plant life, and it has been for billions of years. Coal, oil, and natural gas are the pressed, cooked, and cracked remains of the plants and animals (but mostly plants) that lived hundreds of millions of years ago. All of them—solid (coal), liquid (oil), and gas (natural gas)—qualify as fossils. Because they are also flammable materials and can be burned as fuel, we know them as "fossil fuels."

Oil is mostly burned in the combustion engines of automobiles, coal is mostly burned in order to generate electricity within power plants, and natural gas is mostly burned in the furnaces that power factories. Burning fossil fuel is not the only way to power an engine and to generate electricity and heat, but it is far and away the most common method in use today. Almost 90 percent of the energy used on Earth—to drive, to cook, to light, to warm, to cool, to manufacture—comes from the burning of fossil fuels.

The world's reliance on oil versus coal versus natural gas is split pretty evenly: of all the fossil fuels burned each year, 40 percent is oil, 30 percent is coal, and another 30 percent is natural gas. The reason for this has to do with infrastructure: more than 99.9 percent of motor vehicles were designed to burn refined oil, the majority of the world's power plants were designed to burn coal, and a great part of the equipment in modern factories is powered by natural gas. Switching between fossil fuels is sometimes possible but incurs cost, as the amount of energy released, as well as the type of by-products emitted, differs between fuel types.

As we have seen, the last fifty years is a Story of More—more cars, more driving, more electricity, and more manufacturing; because of this, it should come as no surprise that it is also a Story of More fossil fuel use. During the last five decades, global fossil fuel use has nearly tripled.

Fossil fuels are also known as "nonrenewable" fuels because the amount of time required to transform living tissue into coal, oil, or natural gas is tens of millions of years at a minimum. When we pull oil, coal, and natural gas from the earth and then burn it inside our vehicles, power plants, and factories, it is not being replaced. Because it is easy to envision a world with a growing population translating into a growing use of fossil fuels, we are left wondering: How much fossil fuel is still out there?

Geology is a centuries-old science, and the last hundred years have been devoted largely to locating, mapping, and accessing the rocks that contain oil, coal, and natural gas. Based on this work, the "proven reserves" (that is, the for-sure accessible deposits) of fossil fuels have been tallied and made public by British Petroleum (BP). More could always pop up—most of Venezuela's known oil reserves have been discovered since 1980—but for the most part, the BP databases give a pretty good idea of what's in the ground.

The world's total proven oil reserves constitute a fifty-year supply, given today's rate of oil use; similarly, the total proven reserves of natural gas would also last fifty years if burned at today's rate. The world's proven coal reserves are much larger:

they would take about 150 years to burn through at the current rate of use. Of course, all of those numbers are much too high if fossil fuel use continues to increase each year, as it has done during the last several decades. Inciting panic over fossil fuels running out has never really paid off, and scientists have been accused of crying wolf since 1939, when the director of the Naval Petroleum Reserves informed Congress that America's oil reserves would not outlast a second world war.

It is true that the total amount of known oil and gas reserves has doubled since 1980, but global fossil fuel consumption doubled right along with it. We may not know exactly when, but at some point something will have to give: there is no second Panthalassa Sea that four generations of geologists missed when they mapped the plains, mountains, and ocean floor. If we want human society to outlast the finite resource that it is dependent upon, then any movement away from fossil fuels is a step in the right direction, and one that can't happen too soon.

One thing we don't talk about enough is the lack of overlap between where fossil fuels are found and where they are used, though it has made for decades of international antagonism, to the point where countries are now repurposing their national food supplies in a desperate attempt toward energy independence. The geography is simple, mostly thanks to the final location of the Tethys Ocean. Half of the world's oil and gas reserves can be found within the borders of the nations of the Middle East, with Saudi Arabia as the leader of the pack. In contrast, fully half of the world's oil and gas is used within the nations of the OECD.

The bald discrepancy between those who have and those who must have has made for nearly a century of intrigue between the leaders of Saudi Arabia and the United States, with plenty of supporting roles on each side. As we saw in the last chapter, the economic collaboration of Middle Eastern countries that resulted in the oil embargo of 1973 was enough to destabilize the American automobile industry for a decade or more, not to mention the role that oil interests have played within America's two most recent wars.

Nonetheless, humanity's infatuation with fossil fuels may represent the greatest resource-based love story of all time. And like a long-married couple tottering off on their final journey through infirmity and into oblivion, the concept of divorce is almost beyond imagination.

The United States has met more than 90 percent of its energy needs during the last fifty years by burning fossil fuels. It drives me bonkers that the American conversation about oil, gas, and coal is never about how much we use, only about where we can get more. In today's post-9/11 world, one-third of the total amount of oil America imports still comes from the OPEC nations: exactly the same situation as before the 1973 oil crisis. We are now, and have been for decades, heavily dependent upon the Middle East for our most basic source of fuel.

America has left no stone unturned and no promise unmade in its desperate and luckless search for a way out of dependence on foreign oil—well, except one: burning less fuel. Proposals to move away from oil in favor of domestic sources of coal and natural gas have met with strong resistance to the pollution attendant upon mining and fracking where you live. Objections notwithstanding, the domestic production of natural gas has increased since 2005, while coal production has stagnated at a thirty-year low. The above adjustments are slight in the grand scheme of things, however, because America burns almost exactly the same amount of fossil fuel that it did in 2005.

Even after taking into account the rocks beneath Texas, America is not an oil-rich nation: the proven oil reserves of the United States amount to less than 3 percent of the global total, which is bad news for any nation so thoroughly committed to the automobile. The United States is rich in plenty of other resources, however, and creative efforts to convert them into an ersatz fossil fuel has given rise to what may be the most absurd environmental innovation of the twenty-first century: the conversion of food for people into fuel for automobiles.

Experienced distillers know that anything containing carbohydrate can be fermented into alcohol, given the right setup. This includes sugars such as those from grapes or honey, but it also includes starches such as those within potatoes or barley. Very recently—just since 1990 or so—the edible alcohol "ethanol" has been produced on an industrial scale for the purpose of mixing with, and thus extending, the gasoline supply. At present, ethanol is produced from the fermentation of corn in the United States and from the fermentation of sugarcane in Brazil.

This means that tens of millions of acres around the world are planted, irrigated, and fertilized, pesticides and herbicides are applied, picked, and processed—all for the purpose of mashing and fermenting the harvest into fuel. It's terribly inefficient from a resource standpoint: fossil fuel–powered tractors drive millions of miles and apply tons of chemicals to make it happen, and the whole process has been subsidized to the tune of $40 billion, just to make the farmers' ends meet. The one benefit biofuels confer is that they can be produced domestically and could therefore make the producer less dependent on foreign oil.

To some extent this is working, at least for Brazil, which is able to support itself while importing one-tenth the amount of oil imported by the United States. The United States, on the other hand, imports exactly as much oil as it did in the 1990s, before biofuel technologies really took off. Like most energy innovations of the last one hundred years, development of the corn-ethanol biofuel has simply been used to facilitate increased fuel consumption.

The European Union (EU) has backed itself into a particularly tight corner energy-wise, in that it consists of five hundred million people who have almost zero oil reserves of their own, have largely sworn off coal, but still burn a massive amount of fossil fuel each year. In addition, the EU does not contain vast crop fields that could be fermented into ethanol. Because of this, they have turned to biodiesel.

Diesel fuel is denser than gasoline and is the preferred fuel for heavy transportation engines such as those found in trucks, trains, and military vehicles. Diesel can be extended with oils, such as those extracted from soybean meal and canola. Between them, the countries of the European Union produce five billion liters of biodiesel each year, which equals half the biodiesel production in the world. To sum up, the vast majority of the biofuels produced and used in the world come from just three geographic regions: the United States and its corn-based ethanol, Brazil and its sugarcane-based ethanol, and the European Union's soy-canola-based biodiesel.

Biofuels are considered "renewable" because every year that we renew the world's agriculture, we get the option of taking a portion of the harvest, mutilating it, and then setting it on fire.

Biofuels do not, however, present a realistic alternative to petro-leum at today's level of consumption: if the United States gave up fossil fuels cold turkey and relied 100 percent on biofuels, its current annual biofuel production would last about six days. The story is worse for the European Union: they'd last three days. Brazil could go a little longer—about three weeks, by my calculation. But none of these nations produce enough biofuel to do any more than just take the edge off of their fossil fuel dependency, though they are channeling a large portion of their national agriculture into the effort.

There's an additional ethical consideration with biofuels, in that using them really does amount to torching a massive pile of food inside the engines of many vehicles. It takes more than twenty pounds of sugarcane to make just one pound of biofuel, and the numbers for corn and soybean conversion aren't much better. Today, 20 percent of the grain grown on planet Earth is converted to biofuel—that's a huge fraction for a planet that is also home to eight hundred million starving people, though bio-fuel enthusiasts are quick to point out that the raw material for biofuel production can and sometimes does include both edible and inedible portions of the plant.

We are addicted to our cars, and our cars are addicted to oil, and there seems to be no way out. This intensifies when you dis-cover that today's cars are also mostly *made* of oil. The bumpers, the doors, the dashboard, the casing for much of the engine, and the tires are all made from petroleum-derived "polymers." But it's not just our cars, it's our whole lives that are stuffed with, and packaged in, the synthetic material that we call "plastic"—which is yet another product derived from oil.

At St. Bernard State Park, there is a crawdad-carpeted campground where you can lie awake in the velvet darkness and listen to the alligators bark. Only fifteen miles from the heart of New Orleans, St. Bernard State Park is easy to get to, just follow the Mississippi River as it rolls out toward the Gulf of Mexico. Be prepared to take your time; the speed limit is still fifty-five miles per hour, but most folks go about thirty-five. You'll pass sleeping dogs waiting for sundown and morning glories waiting for sunrise and pawnshops waiting for payday. There's nothing south of New Orleans that hasn't mastered the art of waiting.

Almost exactly halfway to the park, you will enter a bizarre landscape that looks to have been co-designed by Ayn Rand and Aldous Huxley. For five full miles along the road next to Chalmette, Louisiana, you'll pass a massive array of smokestacks, steamstacks, and chimneys, looped together with pipes and shutoff valves, more angular than a Dr. Seuss drawing but equally fantastic. Miles of scaffolding and external stairwells make for climbing trails among the coughing conduits. The stacks are bordered on all sides by huge vats and holding tanks and parking lots where thousands of tanker trucks come and go and come back again. In the daytime, the whole place seems a colossal junkyard of rust and peeling paint, but by night, yellow floodlights bathe it in an eerily forgiving jaundice. A constellation of bright red safety lights delineate the structure to planes and satellites flying above.

The industrial morass that borders the highway at Chalmette is actually many individual factories that all coalesce into a single facility: an oil refinery. There are 135 similar refineries in America, and they are mostly located in Louisiana, Texas, and California.

Each day, the United States imports ten million barrels of

crude oil that came directly out of the ground via an oil well. These ten million barrels, along with another thirteen million barrels produced domestically, must be processed, or "refined," before they become useful. Oil refinement is largely a distillation process: when crude oil is subjected to heat under pressure, it splits into its various parts. The heavier waxy fractions settle, while vapors rise and condense into light fractions such as propane or gasoline, and diesel is left in the middle. The simply built refineries of the 1800s distilled kerosene from crude oil, which was burned in lamps to light, for the first time, the streets of cities and the depths of coal mines and the decks of ships sailing at night. They also produced a variety of flammable by-products that must be useful for something, figured the innovators of the time, and they set forth to find it.

One of the major by-products of any oil refinery is something called "petrochemical feedstock," the raw material from which plastic is made. It's not an exaggeration to say that the invention of twenty or so different types of plastic since 1950— usually recognizable by their "poly-" monikers: polyethylene, polypropylene, polystyrene, polyvinyl chloride, polyester—has revolutionized almost every aspect of how we live.

Before I lecture about plastics, I ask my students to take out a backpack or a purse and count the number of plastic things in it. It is the rare Spartan that cannot find at least twenty items, and it's not unusual for meticulous pupils to tally more than fifty, if they scrutinize the contents of their wallet and separate a ballpoint pen into its working parts. The surfaces that surround us, the fabrics that cover us, the objects that we handle, are to a great extent made of plastics that did not exist when I was born in 1969. Over the short half century that has passed since then, most of the items that used to be made from glass, metal, paper, and cotton have been replaced by plastic, rendering them much lighter, more durable, and far cheaper to produce and transport.

The invention and innovation of plastic truly qualifies as one of the manufacturing miracles of the twentieth century. Plastic surfaces and objects are easily cleaned and kept sterile, leading to their proliferation within hospitals and medical facilities. Plastic film inhibits spoilage and so extends the shelf life of meat and vegetables from days into weeks, allowing for the distribution and consumption of fresh food all year long. Replacing heavy metal parts with lighter molded plastics is a large part of what made our cars, trucks, and planes more fuel efficient during the 1980s, though we offset that advantage by making even bigger vehicles and traveling more miles inside of them.

At present, global production of plastic is more than three hundred million metric tons per year; all put together, that's about equal to the total weight of all persons on Earth. Today's total plastic production is ten times larger than it was in 1969, which had grown quickly up from zero starting in about 1940. Fortunately or unfortunately, the majority of plastic produced each year is consumed as disposable packaging. On average, a person living within the OECD discards just over his or her body weight in plastic every single year, and despite burgeoning efforts to recycle, more than 90 percent of it is simply dumped into the landfill. Almost 10 percent of the plastic that we throw away eventually finds its way out to sea, where it has congregated into massive floating rafts of trash that perpetually swirl atop the oceanic gyres.

Almost all of the world's plastic is manufactured from oil, and oil is used to fuel the factories where it is produced. Making plastic accounts for 10 percent of all the fossil fuel burned on planet Earth every single year.

When we talk about energy, be it fossil fuels or renewables, it is easy to get tangled up between percentages and total amounts, and I've seen both politicians and scientists engage in this sleight of hand. It goes like this: My friend Brian quit smoking a few years ago, which was quite a feat because he had been addicted for decades. At age sixteen, he smoked after school with his friends, and a pack of cigarettes would last him about a week. While at community college, he got a part-time job and so upped his habit to about two packs a week. After graduating, he got a full-time job on a construction site, and soon thereafter he was smoking a pack a day.

When I want to minimize the significance of cigarettes in Brian's life, I emphasize that the percentage of his salary that he devoted to the purchase of cigarettes decreased dramatically during the twenty years that passed. When I want to maximize the significance of cigarettes in Brian's life, I emphasize that the total number of cigarettes he smoked each week increased by a factor of seven over that same time period. Both of these statements are factually correct, but when they are presented in isolation, they tend to give different impressions of Brian's habit. To fully understand the role that cigarettes played in Brian's life over those formative years, it is best to understand both trends.

The peoples of the earth, far and wide, are divesting themselves of fossil fuels and have been since the 1960s. What this actually means is that fifty years ago, the global population met 94 percent of its energy needs using fossil fuels, and ever since then, that percentage has been decreasing; today, it is down to 85 percent. This is true even in the United States and especially within Europe. I have often seen this set of *true* facts presented as evidence that we are moving in the right direction.

It is also true, however, that the total amount of fossil fuels we burn has increased substantially over that same time period. Globally, the total amount of fossil fuels burned each year has more than doubled over the last fifty years, while the total amount burned in both the United States and Europe has increased by one-third. When we focus on the total amount, we get a clear picture of a planet that is burning more and more fossil fuels every year, but when we put the two sets of information together, as a good scientist should, we begin to see the real picture.

We are using more and more energy, but we have begun to find other ways to get it besides fossil fuels. The magnitude of these "other ways" is no more than a layer of icing on top of a bigger and bigger piece of energy cake that we eat each day. What we never seem to do is stop and ask ourselves if we really need dessert. We've put off that question again and again, while the refineries pulse and the world burns.

13

The Wheels We Turn

*He that abstains
To help the rolling wheels of this great world,
Glutting his idle sense, lives a lost life,
Shameful and vain.*
                    —*The Bhagavad Gita* (c. 200 B.C.)

A spinning wheel is a magical thing that moves and stays put at the same time. If you wrap copper wire around an iron stick and attach it to a wheel, you can make electricity by spinning the wheel near a magnet. This transformation of mechanical energy (the spinning wheel) into electrical energy (the current in the wire) is a type of sorcery that has inspired curiosity since it was first discovered two hundred years ago. So far, the twenty-first century has been, if nothing else, a joyful new experiment with the spinning wheel.

Everyone has a favorite city, be it Paris in the spring, Cairo at sunset, or Christmas in New York. Mine is Minneapolis during July, where people suspend their summer labors to fully enjoy the sunshine. Minneapolis was both the first place where I moved after I left home and the first town that I lived in that had more than one elevator, so I will always associate the city with glamour and freedom and multiple opportunities to rise.

My favorite place to go during summer in Minneapolis is the Stone Arch Bridge. It is a huge granite-and-limestone plaza that spans the Mississippi, linking downtown to the University of Minnesota. You can walk, bike, or skate across the bridge, loiter on its plaza and take photographs, or simply sit and people-watch as the river goes by. The midpoint of the Stone Arch Bridge is the perfect place to stop and think about life for a while on a sunny July day.

There's another reason why the bridge is wonderful on a hot day: the spray coming off the rushing water keeps you nice and cool. On the upriver side of the bridge gushes the Saint Anthony Falls, emitting a continuous low rumble in the background. So much water, endlessly falling, is just a brief stumble along the Mississippi River's two-thousand-mile journey into the Gulf of Mexico. No matter how long you stand mesmerized by the Saint Anthony Falls, you won't see everything, for deep underneath the water's surface, in the darkness, turn five colossal wheels.

Each of these enormous wheels is generating electricity, using the wire-and-magnet method described earlier. As the river falls fourteen feet down over the dam, it turns five big wheels called "turbines." Over the course of 2016, the Saint Anthony Falls generated almost a hundred gigawatt hours of electricity— enough electricity to power the city of Minneapolis for a week. There are hundreds of similar "hydroelectric" facilities across

America, and thousands around the world, and the basic principle is the same in each one: Dam a river, let the water fall down over a set of turbines, then harvest the electricity made by the giant turning wheels.

The largest hydroelectric plants turn more than thirty turbines and generate up to two hundred times more electricity than the small but respectable Saint Anthony Falls facility in Minneapolis. Hydroelectricity is a powerful source of energy, and it is by far the most extensively used alternative to fossil fuels the world over. Even so, hydroelectric power plants generate only 18 percent of the world's electricity.

But water isn't the only thing that can turn a wheel. If you drive south from Minneapolis toward San Antonio, you'll see multiple clusters of white wind turbines that have sprung up like dandelions during the last ten years. A strange metal forest of angular trees, their bony branches are alternately turning or frozen. Each blade of these gigantic white fans is the size of an airplane wing mounted two hundred feet off the ground. When one of them catches the breeze, the spinning generates electricity, just like a turbine underwater. The electrical current then flows into our homes, buildings, and factories via a grid of above- and below-ground cables.

Wind farms contain hundreds of carefully spaced turbines all working together; such facilities regularly produce about twice the amount of electricity produced by the Saint Anthony Falls hydroelectric station. But compare the number of turbines in the two types of facilities—a few underwater versus a few hundred in the wind—and you'll get a sense of the power of water over air. The total wind-powered electricity generated in a year amounts to less than 4 percent of the total electricity that is consumed around the world.

Solar power can turn a wheel in particularly sunny places. For this, a series of mirrors and lenses is arranged to concentrate and focus sunshine so intensely that it boils water into steam. As the steam ascends toward the sky, it blasts by turbines, turning them to generate electricity. This concentrated solar power, combined with the electricity that can be produced between the glassy layers of solar panels, accounts for less than 1 percent of all the electricity generated each year on planet Earth.

The electricity generated by solar and wind power is considered green, clean, renewable, and environmentally friendly because no by-products are emitted and for as long as the wind continues to blow and the sun continues to shine, the turbines will continue to spin—that is, no resources are noticeably used up and no ecosystems are disturbed during the process. It would be great if we could also harness the energy that we use to *talk* about renewables, for we tend to do it in gross disproportion to the amount of power they generate: wind and solar power, put together, provide less than 5 percent of the electricity used on planet Earth.

There is one more essentially limitless source for the generation of electricity, but it produces one hell of a by-product. Naturally radioactive metals, such as uranium, spontaneously decay and emit energy while they do it. Deep inside their concrete core, nuclear power plants incite decay within specially "enriched" uranium and extra-large amounts of energy are released. This activates a chain reaction, in which the products of one decay event trigger the next decay event, and so on. The total energy released is used to boil water into superheated steam, which rises past a turbine and generates electricity. Nuclear power plants are heavily engineered to modify the speed of the chain reaction, which is important, because a runaway nuclear reaction is not unlike the detonation of a nuclear bomb.

Eventually, under normal functioning, the enriched uranium source will decay to the point where it is no longer useful for generating electricity and must be discarded as waste. This is tricky, because while spent uranium does not emit enough energy to turn a turbine, it still emits more than enough to be a hazard to plants and animals and, of course, humans. The specter of a horrific engineering failure, combined with the extreme toxicity of the waste produced, has led people to both fear and shun nuclear power.

When nuclear energy debuted in the 1970s, it was met with even more optimism than we see for wind and solar power today, only perhaps better placed: an average-sized nuclear power plant generates the equivalent of eighty Saint Anthony Falls hydroelectric facilities, more than enough to power a city of half a million people for an entire year. A series of disasters, including Three Mile Island in 1979 and Chernobyl in 1986, alerted us to the fact that nuclear power plants—like all technologies—are only as competent as whichever Homer Simpson is at the helm.

In addition, the tools and skills required to enrich uranium for use in generating electricity are part and parcel of those required for constructing nuclear weapons, which has created a geopolitical disincentive for widespread adoption. For all of these reasons and more, nuclear power has been on the decline since the early 2000s. The global fraction of electricity generated by nuclear facilities has decreased from its all-time high of 6 percent in 2002, and several European countries have announced plans to decommission the nuclear power plants that remain.

The United States, by comparison, is still heavily invested

in nuclear energy. Almost 20 percent of the electricity produced in the United States is generated by the country's one hundred or so nuclear power plants. America's reliance on nuclear energy is about four times as high as the global average, and the amount of electricity used in the United States that comes from nuclear power plants is nearly double the amount that comes from hydro-, wind, and solar power combined. For all the drawbacks associated with nuclear energy, there is one important by-product that it does *not* produce: carbon dioxide. You'll see why that's important in the next chapter, but for now, let's stick with spinning turbines.

Renewable energy is popular—it is often touted as the "fastest-growing" source of energy with statements such as "the total amount of electricity harnessed from wind has more than doubled since 2010" and "the rate of solar power installation has grown by a factor of more than one hundred during the last ten years." Here we see my friend Brian and his cigarettes all over again: true statements that mask the fact that, taken together, wind and solar still provide less than 5 percent of the world's energy diet, and it's terribly hard to see a way to bring this up to even 50 percent. No, the vast majority of the turbines in the world are turned by the burning of fossil fuels.

Power plants burn coal and natural gas in order to boil water into steam that rises past the turbines, making the great wheels spin; alternatively, gases produced by burning fuel simply stream through turbines specially designed to function at great heat. Two-thirds of the world's electricity is made by turning turbines with fossil fuels; in the United States, this proportion is slightly higher. In fact, there are only a handful of countries in the world that do *not* generate the majority of their electricity using fossil fuels. Norway is one example, and it can get away with it only by being a landscape lavishly endowed with seasonally pulsing mountainous waterways and home to a tiny population.

Electricity was indeed a magical boon, and its discovery forever changed the way that human beings relate to the objects around them. Unfortunately, the conversion of mechanical energy into electrical energy is inescapably and depressingly inefficient. You've probably noticed this if you've ever ridden one of those bicycles at the science museum, the ones that are attached to a single light bulb. You have to pedal like gangbusters to get even a dim flicker started, and the instant you slow down, it goes out.

The best way to compensate for this inefficiency is to start with a very dense energy source. Fossil fuels and enriched uranium store huge amounts of energy, but a summer breeze or a sunbeam? Not nearly as much. A rushing river is somewhere in the middle. At the current rates of electricity consumption, a modest-sized U.S. city of one hundred thousand people—and there are hundreds of such cities in America—would require the full-time operation of one average-sized power plant fueled by coal.

You may have heard the stupendous, nearly endless, amount of total energy in the sun quoted, or the energy that causes Earth's air to move in great patterns, but in reality, windmills and solar panels capture a miniscule amount of these totals. Powering an American city with a population of one hundred thousand people using renewable energy would be truly impossible, unless there happened to be a massive waterfall nearby. Otherwise, ten medium-sized hydropower facilities would be needed. Alternatively, one thousand wind turbines could do the job, or one million solar panels, just to keep the lights on for one hundred thousand people. Of course, all of this assumes that the city is located in a sufficiently wet, windy, or sunny area, as it is based on electrical generation under near-optimal conditions.

Switching whole hog to renewable energy at today's level of electricity consumption is not possible for the United States, although I've heard people talk about it as if it were an attainable goal. At today's rates of electricity consumption and generation, powering America using only hydropower would require fifty functioning Hoover Dams within each one of the fifty states of the Union. Powering America using only wind power would require more than one million wind turbines, or one every mile or so across the whole of the continental United States. As for solar energy, a land area the size of South Carolina would have to be sacrificed to solar panels in order to generate America's annual diet of electricity. Entirely switching over to renewables at their present rate of efficiency is, unfortunately, a pipe dream.

There are some who prophesy that today's enthusiasm for green technologies will naturally translate into improved efficiency and the ability to scale. But we'd need a breakthrough that could improve the output of energy by an order of magnitude. Such ingenuity might also be usefully applied to improving the safety and security of the mere five hundred nuclear power plants that would be needed to supply all of America's electricity.

I do believe that renewable energy is part of the *Use Less and Share More* solution and that there is room to meet in the middle of using less electricity while making a larger fraction of it from water, wind, and sun. However, there's an additional issue associated with scaling up renewables that has to do with where the necessary metals are going to come from. At present, a large part of the cadmium, copper, lead, tellurium, zinc, and lithium used in the turbines that create electricity and in the batteries that store it come from just two countries: Chile and Peru. Historically poor countries, Chile and Peru have remained poor, even as the demand for these metals has skyrocketed in keeping with the sales of battery-powered devices such as laptops and cell phones.

Other sources might be secured from the mines of Australia, or even Kazakhstan, but the United States will not be able to supply what it needs of these metals from domestic sources. Similar to the problem with fossil fuels is that lithium and cadmium deposits are not evenly spaced throughout the world, and the countries that are becoming dependent upon these metals are not the same countries that can produce them.

Most of the Americans I meet don't realize that their iPhone runs off of fossil fuel. When you charge your laptop or phone, you pull electrical current out of the wall, and that current was more than likely generated by a coal-burning power plant located on the outskirts of town. Your refrigerator, toaster, and television, and all the electric lights in your house, work this same way, and it's most often the case that they all go back to the burning of fossil fuels, as does the electricity that lights your school, hospital, and workplace and also powers the machines within them. This is true as well for the electric car that you sometimes think about buying. Tethered to fossil fuels through an umbilical cord made of lead, nickel, cadmium, or lithium, an electric vehicle emits its smog on the other side of town, allowing us to imagine it as clean and green.

If nothing else, fossil fuel use has gotten a lot cleaner over the last five decades. Big steps have been taken to scrub much of the lead and sulfur out of the emissions that are produced during the burning of fossil fuels, and air quality has increased dramatically in many large cities as a result. However, the biggest pollutant coming from our power plants and vehicles is something that you can't see or smell, and you might not even notice it becoming a problem.

It's a gas called carbon dioxide, and there's more and more of it every year, and it just might kill us all.

Part Four

EARTH

Attachment to matter gives rise to passion against nature.
—The Gospel of Miriam of Magdala (c. A.D. 150)

14

Altered Air

*In the last few decades entire new categories of waste
have come to plague and menace the American scene.
These are the technological wastes—the by-products of
growth, industry, agriculture, and science.*
— President Lyndon B. Johnson (1965)

Every person I've asked has offered me a unique definition of our
most universal concept: the circle of life. A friend from childhood
told me how her first granddaughter was born on the same day
that her own grandmother died and that the equal weight of joy
and sadness combined to convince her that life cannot be lost or
gained, only lived and loved. A stranger on the A train busted out
with the Disney song when I asked him, crooning from memory
the blockbuster melody that Elton John wrote in less than one
hour. As for me, I automatically think of plants when I hear a
reference to the circle of life: specifically, of how energy goes in
when plants grow and how it comes back out when plants burn.

Studying biology is like studying a Hieronymus Bosch

painting: the chaos that you sense from a few steps back only increases as you lean forward to examine it more closely. It's not only the endless shapes and textures of the world's plants and animals that overwhelm, it's also the endless variation of how these living beings grow, maintain their parts, store against hard times, defend against their enemies, and reproduce yet another generation. In nature, as within Bosch, the more you look, the more you see.

There's really only one thing that all of Earth's life-forms have in common: their insides burn. Every living organism, from the single-celled microbe to the cheerful daisy flower to the hundred-ton whale, is capable of burning plant tissues (yes, plants can burn their own tissues!). The human body is very good at burning plants: we harvest energy from the breakdown of sugars, proteins, and fats within the plants that we eat or within the animals that we eat (animals who ate plants, after all). We use this energy to fuel our bodies, to run, walk, talk, think, breathe, and everything else that we do.

Plant cells—and only plant cells—are capable of doing the reverse. Plants need, first and foremost, two things in order to live: energy and carbon. Plants get energy from the sun in the form of light and they get carbon from the air in the form of a gas called carbon dioxide. When we burn food in our bodies, we use the energy to live and we exhale carbon dioxide out of our lungs. Thus, the history of life is essentially a great gala with Earth's atmosphere functioning as the main ballroom. The dance is a two-step: energy and carbon dioxide in, energy and carbon dioxide out.

Inside a tummy is not the only place that a plant can burn. When wood burns in a fireplace, the energy that was originally collected from the sun is released as warmth, and carbon dioxide drifts up and out of the chimney. When we burn fossil fuels—that is, long-dead plants—inside our car engines, energy harvested from the sun millions of years ago runs the engine, and carbon dioxide sequestered millions of years ago is released into the air.

During the last half century of mining and then burning fossil fuels, humans have released a huge amount of carbon dioxide into the sky that would have remained buried deep in the earth otherwise. Since 1969, the nations of the globe have burned enough coal to fill a grave the size of Texas and a volume of oil large enough to fill Lake Pontchartrain three times over. While we used the energy to run countless machines, we also dumped more than one trillion tons of leftover carbon dioxide into the air.

Increases in carbon dioxide are easy to measure—there's a weather station in Hawaii that does it every day—and recognizing the increasing trend in the data is as obvious as recognizing your own name written on a chalkboard. Every single scientist I know is freaked out by the steep increase in carbon dioxide of the last fifty years. But we are more freaked out by the fact that our governments are not as freaked out about it as we are.

We have a saying at my job: "Six months in the laboratory can save you an hour in the library."

Because I research plant biology, I am very interested in the amount of carbon dioxide that's in the air right now and how any increase might affect plant life. After all, the plants around us are more than a decoration—they are food, medicine, and wood: three things that human civilization cannot do without. As the burning of fossil fuels increases the amount of carbon available to plants, there is every reason to believe that agriculture, pharmacy, and forestry could be affected.

During 1999, my team and I designed and built four growth chambers—Plexiglas boxes made using materials from Radio-Shack and Home Depot. They work great and look quite elegant; we can finely control how much light, water, and carbon dioxide our plants have to work with. These days, we routinely grow hundreds of plants from seed to flower, monitoring them every few minutes for weeks on end. For the first few years, however, getting the chambers to work at all required lots of trial and error. Keeping them cool turned out to be the biggest challenge.

We designed our chambers to be able to hold large amounts of carbon dioxide gas inside them. While the plants grow, we run a stream of carbon dioxide over them, then exhaust it out through the ventilation system of the building. We need the air to move: we don't want our chambers to become foggy terrariums, we want our plants to feel as if they are outside in a farm field. After we raise carbon dioxide to the right level, we turn on the artificial sunshine. Early on, when we fired up the chambers and walked away, we returned to find that the temperature inside the chambers had shot up past ninety degrees Fahrenheit within an hour of turning the chambers on. We were puzzled: this had all happened inside a building kept cool at sixty-five degrees.

We tried everything to cool our chambers down. We set the lights as low as possible, we blocked all the windows to keep outside sun from coming in, but none of it made any difference. Even after the lights turned off and the room chilled down, the temperature inside these chambers remained high. In contrast, our "control" chambers—the ones with no extra carbon dioxide running through them—did not heat up significantly after the lights turned on; they stayed at a cool seventy-five degrees or so.

I should have anticipated this. Other scientists had seen the same thing as early as 1856, when Eunice Foote described how glass jars that she filled with extra carbon dioxide not only became "much heated" in the sun but were "many times as long in cooling." A few years later, John Tyndall constructed an elegant brass device that could be filled with a known amount of carbon dioxide and then subjected to heat; in the end he came up with the same result as Eunice Foote, except he was the one who got the credit for it.

In short, the carbon dioxide molecule has a unique shape that intercepts and then absorbs heat. Add just a little carbon dioxide to a chamber of air, then let the sun shine through it, and the whole thing will heat up much more than a chamber without the extra carbon dioxide. This simple fact has been in our chemistry textbooks for more than one hundred years, but it still took me six months to understand it when I saw it with my own eyes. How slowly grinds the whetstone that sharpens the human mind.

Scientists have also been trying to get politicians to act on this information for more than one hundred years. As early as 1896, Svante Arrhenius warned that fossil fuel burning would cause global warming. Between then and now, the carbon dioxide content of our atmosphere increased by more than one-third. So, the earth should have gotten warmer, right?

It did and it didn't. But mostly it did.

15

Warming Weather

Feel by turns the bitter change
Of fierce extremes, extremes by change more fierce.
                                        —John Milton (1667)

Whenever I think of weather, I think of wind. I think of the balmy breath that wafts down the Mānoa Valley in Hawaii, of the searing gusts that hit your face like a hair dryer in the Nevada desert, of the bracing breeze that washes over the Oslo Fjord, and the chilling blast that greets you just past the Minneapolis airport's sliding doors. I think especially about wind that carries water, about the way that the North Dakota sky explodes during a summer thunderstorm, of the horizontal gales of Puerto Rico, of the dumping snow that buries Anchorage in a matter of hours, of the sizzling glare that pulls the moisture from your picnic plate in Arizona, of the chilly mist that makes your joints ache when you so much as look out the window in New Zealand.

Wind is, of course, moving air. Whenever air moves, it is because something is pushing it or because something is pulling it. You've heard the weather report mention high or low pressure systems when there's a storm brewing—that's because air moves from an area of high pressure to an area of lower pressure, creating the winds that move the storm. The energy that fuels the movement of wind ultimately comes from the same place where most energy comes from: the sun.

What would it take to make a planet with no wind and no weather? First and foremost, we'd need to turn off the sun. Because the sun shines more directly on the equator than on the poles, and because of the differences in how the sun warms water versus rock versus snow, the temperature of the earth's surface varies from place to place. The sun warms our atmosphere from above, while the surface warms it from below, creating pockets of cold air in the middle that sink and displace the warm air that rises. Every moving parcel of air pushes some other parcel out of the way, so the beating of a single butterfly wing on one side of the world adds to the breeze that sways through the treetops a thousand miles away. Sunshine also helps pull water vapor up from the ocean into the sky, where it comes back together and falls as rain and snow. For all of these reasons, getting rid of the sun would be the first step in getting rid of weather.

Once the sun was turned off, we'd also need to stop the earth's rotation, as the constant turning of the globe tweaks the winds such that they draw moisture up from the Sahara and rain it back down on the Brazilian rain forest. As a final step, we must also wait several billion years for the earth to cool from the inside out and lose every last bit of the heat from its molten babyhood and radioactive core, for these are what drive the ancient cough of the volcanoes. Until that far-off dawnless day, after our favorite star has fully flickered out and the planet wakes cold and still and dark, there will be both wind and weather.

Because the sun is the primary source of the energy that drives weather, and because of the carbon dioxide molecule's well-understood ability to absorb sunlight, the sensible notion that Earth will get warmer as carbon dioxide levels rise has been dubbed "the greenhouse effect," calling to mind the feeling of being in an unusually and artificially warm place.

It may shock you to hear it from me, but I appreciate how the idea of a greenhouse effect—or at least a thoroughly sinister one—can be a hard swallow.

Those of us who grew up *cold*—who shivered at the bus stop after closing time, who pushed the old Ford out of the ditch while wearing pantyhose, who walked the back forty for hours looking to see where the horse got out, and who did all of these things in a thirty-below windchill—are going to be very hard to convince that one-point-five degrees of warming is a categorically bad thing, and I don't think that the Intergovernmental Panel on Climate Change has fully wrapped its collective head around that challenge.

I am so often accused of exaggerating the cold climate of my childhood that I recently printed out and laminated the National Weather Service records for 1975, the year that I started grade school. That was the year that a four-day blizzard commenced on January 9 and has since become known as the record-breaking "Storm of the Century." During that terrible storm, a full year's worth of snow came down in one week. In fact, 1977 to 1979—the formative years of my skating, sledding, and snowman-building career—was a three-year spate of the absolute coldest winter temperatures ever recorded in the contiguous forty-eight states during the twentieth century.

So yes, it was cold, and we knew it was cold, and we were used to the cold, and we did not like the cold, and it especially sucked when our mittens were wet. But at least I *had* mittens, my mother is always quick to remind me, for she did not during the almost-as-cold winter of 1936, which is why the very tip of her nose is snub and not pointed like mine.

Minnesota, along with the rest of the Midwest, has become downright balmy since I was a kid, and more and more of what used to be snow now comes down as rain. The season when there's snow on the ground is two full weeks shorter than it was in 1972; the ice covering Lake Superior thaws three weeks earlier than it did in 1905. For a whole bunch of reasons, including too-thin ice, kids don't skate on the ponds in my hometown anymore, but they do sneeze out ragweed pollen for an additional fifteen days compared with when I was a kid. There's nothing like visiting your old childhood haunts to sharpen your perception of how passing time has changed the world.

In addition, the effects of rising carbon dioxide are not as simple and uniform as the "greenhouse" moniker suggests. As carbon dioxide absorbs infrared radiation, hot pockets of air get hotter. This causes the contrast between warm and cold pockets to get bigger and incites faster evaporation from the ocean, which puts more moisture into the air that eventually has to fall back out somewhere. Because of this, scientists have been predicting an increase in the frequency and intensity of big storms since the 1980s. Their idea of "global weirding" makes sense; as I write this, it's a petrified five degrees Fahrenheit here in Oslo and a slurpy forty-three degrees Fahrenheit two thousand miles north of here, in northern Greenland. There are plenty of reasons to suspect that the massive hurricanes, the punishing snowstorms, the drenching downpours, the crazy reverse freezes, and the ruthless droughts of the last two decades were exacerbated by the extra energy trapped in our atmosphere colliding with normal weather systems and then whipping them into a frenzy.

Still, two centuries of burning fossil fuel, and emitting carbon dioxide into the atmosphere, where it absorbs more of the sun's energy, should have raised the overall temperature, right? Yes, it should have, and yes, it did. The average global surface temperature has increased more than 1.5 degrees Fahrenheit during the last one hundred years (that's a little short of 1 degree warming Celsius).

I cannot overemphasize how clear this data is. The instrument used to measure temperature rise—the thermometer—is foolproof: it was invented three hundred years ago and perfected soon after. Weather stations that record the temperature are everywhere, even in the most remote localities of the world, and the data has been faithfully retrieved and recorded in detail for centuries by devoted farmers, postmasters, nuns, citizens, and scientists. The most recent warming trend of just over one degree Fahrenheit during the last thirty years is not something that scientists fight about—and trust me: scientists will fight about almost anything.

The ten-year period between 2005 and 2016 was the warmest decade on record since the invention of the thermometer, placing us clearly at the warmest high of a one-hundred-year-long temperature increase. *Somebody should do something,* you might be saying to yourself right about now.

For more than twenty years, the United Nations has been trying to address climate change, and the nations of the world have participated enthusiastically. The first such UN Framework

Convention on Climate Change (UNFCCC) convention took place in 1995 and involved more than 150 different countries, including all the big players: United States, China, Brazil, India, Russia, and Saudi Arabia. They had at their disposal the 1992 report of the Intergovernmental Panel on Climate Change, a now famous document that delineated six discrete climate change "scenarios"—six different scientific projections for the year 2100 that ranged from under two to over four degrees Celsius of global warming (that's three to eight degrees Fahrenheit). The purpose of the UNFCCC was to propose widespread changes in energy use and to get the countries of the world to agree to them, ideally under some kind of binding contract.

Three years later, the UNFCCC produced the Kyoto Protocol, an agreement between industrialized countries and the rest of the world to reduce carbon dioxide emissions back to below 1990 levels. The United States signed the protocol but never ratified it, and Canada withdrew once it became clear that the decrease was not happening. Both the European Union and Russia signed the agreement and then allowed their carbon dioxide emissions to rise regardless. Emissions also increased in China, India, Indonesia, Brazil, Japan, and several other countries that signed but never specified a target for reduction. It was a very nice idea that everyone basically agreed to and then completely blew off.

Seventeen years and four additional reports later, the 2015 UNFCCC produced the Paris Agreement, a climate accord that pleaded for help in limiting net global warming to two degrees Celsius by whatever means available. Every country signing the Paris Agreement was free to chart its own course toward emissions reduction, be it movement toward green technology, reforestation, or simple austerity with respect to fossil fuels. In the end, it was the same old story as before: 175 countries signed, carbon dioxide emissions continued to increase, and 2016 was the hottest year on record.

Nevertheless, the world places great symbolic value on the appearance of loyalty to these doomed protocols. Donald Trump announcing that the United States will not comply with the terms of the Paris Agreement is like me announcing that I will not rule England after Queen Elizabeth dies, but the international media still reported it as news.

The problem is that there's no way to enforce a cultural aspiration posing as an international agreement; so far, nations have gone to war in order to secure fossil fuels, not to desecure them. There's also no short-term economic incentive for stuffing the fossil fuel genie back into its bottle, or for even trying. Fiscal cycles turn much faster than biogeochemical cycles, and there is no industrial profit to be made in pushing a Story of Less. These realities, taken in conjunction with our trajectory of *more*, leads me to sorrowfully conclude that the worst-case scenario projected by the Intergovernmental Panel on Climate Change is not unlikely and may be unavoidable.

You may have heard injunctions to keep the average global temperature rise "well below 2 degrees Celsius"—a phrase taken verbatim from the Paris Agreement. Based on an earnest study of the data, scientists have predicted all kinds of calamity at two degrees Celsius warming—including catastrophic heat waves, droughts, sea-level rise, ocean acidification and crop failures, and a lot of other things—predictions that could be both right and wrong at the same time. Nonetheless, the specter of devastation is frightening, and it seems that every new study serves to make the picture worse.

One puzzling response has been to suggest that the answer to our fear is to become *more* afraid and that our real problem is that we're not terrified enough. "People Need to Be Scared About Climate Change" and "Time to Panic: . . . And fear may be the only thing that saves us," shriek today's headlines, as if from a mirror-universe version of Franklin D. Roosevelt's First Inaugural Address.

Again, I'm just a lab girl, but the idea of scaring the public for the sake of scaring it scares me. People don't make good decisions out of fear, history seems to have shown, and at least some of the time, people who are afraid are also prone to doing nothing. Case in point: When I was a child in the 1980s, I was terrified of a nuclear war. It did not comfort me to see the adults around me further scaring one another (and, by extension, me). Was our fear justified? Yes. Did it yield a solution? Not really. Forty years later, the type of bomb that was dropped on Hiroshima is now used as the matchstick that lights the thermonuclear bomb of today. Despite the marring of my childhood sleep, we now stand capable of more nuclear devastation, not less, at the hand of more nations, not fewer.

We should also remember that the two degrees Celsius limit stems from a question that scientists asked during the 1970s: How much warming could be expected if carbon dioxide doubled? The answer that they came to was two degrees Celsius. Their result spurred other scientists to study the effect of two degrees Celsius warming on weather, agriculture, and society, and they concluded that each sector would suffer greatly. Avoiding this suffering would require putting a stop to the warming before it got to the level they had studied. The two degrees Celsius warming limit has therefore legitimately made its way into countless international reports and online articles.

This also refers to a *net* limit: a recommendation that the *total* warming since the industrial revolution be less than two degrees Celsius. Earlier in this chapter I told you that the average global surface temperature has increased by just under one degree Celsius since the Industrial Revolution. This means that, as of right now, we've got just over one degree of warming before we are within the territory of catastrophe first predicted by the scientists of the 1970s.

My own goal is to inform you, not to scare you, because teaching has taught me to know and respect the difference. I've found that fear makes us turn away from an issue, whereas information draws us in. Thinking logically, in light of the above, it is clear that to avoid warming and upheaval beyond what we've already experienced will require a *transformative* approach to energy use, rather than the incremental changes supplicated within the Kyoto Protocol and the Paris Agreement. We need to transform our collective idea of what energy is *for*, and then transform our individual—and eventually collective—practices of how energy is *used*.

As a scientist, I am supposed to have answers, but like everyone else I know, I cannot yet imagine what real transformation looks like. Almost everything I do uses energy. Almost all of that energy comes from fossil fuels. I am one of one billion people who live like this. I am one of seven billion people who want to live like this.

If the Story of More becomes the Story of Everyone—if everyone on our planet adopted an American lifestyle—global emissions of carbon dioxide would be more than four times what they are today. It is not certain how much of that carbon dioxide would dissolve into the ocean (and how much damage that would do), but as the scientists of the 1970s found, it is very difficult to envision an atmosphere with less than double today's carbon dioxide content by the year 2200. At least two degrees Celsius of net temperature increase will accompany that rise in carbon dioxide, as well as the cataclysms attendant upon that amount of warming.

It's not time to panic, it's not time to give up—but it is time to get serious.

The year 2200 is not so far off, just under two centuries from now. Two centuries ago was when we began to use the steam engine in earnest, mining and burning the coal that started this whole mess. Two centuries from now, two centuries ago—four centuries from the beginning to the end, perhaps to *our* end.

Fate has placed you and me squarely at the crossroads of environmental history.

The year 2200 will also be during the lifetime of our great-great-great-grandchildren. This means that we're down to about three generations to figure out how humanity might survive civilization.

There's a lot more to talk about, and most of it is much simpler than the idea of global warming and weirding, but the stakes are just as high. If you want to get a head start, call up a Canadian during February and ask her what happens when the earth warms. The odds are that she won't mention an increase in hurricanes and forest fires or the expansion of drenching rainstorms and baking heat waves. She probably won't go into the possible effects on crop yields and rates of malaria contagion either. Given the season and the location, Canadians are much more likely to observe (very politely) that when the temperature goes up, all the ice surrounding them starts to melt.

16

Melting Ice

*The floe has been a good friend to us, but it is reaching
the end of its journey, and it is liable at any time now
to break up and fling us into the unplumbed sea.*
—Ernest Shackleton (1915)

There was a while there, when I was six years old, that my best
friend was a piece of ice. It was a hard, thick chunk of ice about as
big as two bricks put together. I named it "Covington."

In Minnesota, winter is like an indigent uncle who turns
up on your doorstep every November. His initial presence light-
ens everything up, gets you outside, and reminds you how great
it is to simply fly down a hill or twirl on one foot. Then January
arrives and your uncle changes into that great-aunt who never
liked your mother anyway. It gets *cold,* and everything ices up
way past what you remember from last year. By the time school
reconvenes after Christmas, the snow has petrified such that

the sidewalk amounts to little more than a rut between two snowbanks.

Us kids walked to kindergarten in the seventies, and though my journey wasn't a long one, it fostered my first thrill of independence as I myself chose which cracks to step on and which ones not to. Early in 1976, I took to kicking a piece of loose ice as I walked, sending it ringing out in front of me every few steps— the frozen North's version of kick the can. One day I arrived at the school grounds with reluctance, having found the perfect piece of kicking ice, its heft and form ideal for my purposes. As the bell rang, I stashed the ice in a snowbank near the fire hydrant. It was still there after school, so I dug it out and kicked it back home, where I left it next to our back door and went inside. I got up the next morning and did it all over again, and Covington was born.

The earth's water is unevenly divided into salty versus fresh. If you took all the water on our planet and shrank it down to fit in a bucket, that bucket would contain one gallon, thick with salt, of sloshing ocean. The amount of fresh water, by comparison, would be three spoonfuls. Of those three spoonfuls, two of them would be frozen into ice.

The global temperature rise that we discussed in chapter 15 is melting the world's ice. Scientists are certain of this, because we see it using satellites at work, we see it in our daily lives at play, and we see it through our binoculars while on vacation.

Nowadays, the number of days below freezing in Canada don't last the whole season of any respectable-sized Pee-Wee League, so boys and girls no longer play pickup games of hockey on neighborhood ponds. For the same reasons, an ice-skating marathon in the Netherlands hasn't been held for twenty years. Around the world, whole generations have grown up without ties to something the previous generations cherished, owing to too-thin and too-brief ice.

Shuttled into climate-controlled facilities, the most talented of these young athletes dream of the Big Rink. Should some of them make it to the Winter Olympics one day, it is not unlikely that all of the sports will have to be played inside: of the twenty-three sites that have hosted the Winter Olympic Games since 1926, almost half can no longer accommodate the skiing, skating, and snowboarding conditions for the games today. This includes the city where I live, Oslo, Norway, which hosted the Olympic bobsled event just outside my north-facing window, in 1952.

Even we amateur skiers are running out of options quickly. Since 1930, Montana, Utah, and California have lost one-fourth of their snowpack to melting. In Colorado it's much worse: some areas have lost more than 80 percent. Perhaps the United States will follow in the footprints of Germany, Finland, and Norway, where indoor ski resorts are being built (at a wildly high energy cost) in order to preserve some version of snow-play for our grandchildren.

The sculptured ice of Glacier National Park in northwestern Montana has been making tourists swoon since the park opened in 1910, but if you want to see it for yourself, take my advice and don't put it off. Take your family with you, for all of the glaciers in the park are on track to disappear within our kids' generation. The regal glaciers of ice that rest across the mountains of Switzerland, Alaska, New Zealand, Tanzania, and Norway (just to name a few) are melting at a furious rate. On average, the earth's glaciers have been losing ice since at least the 1970s, and this melting trend has accelerated over the last ten years. The average glacier in the Rocky Mountains has melted by half since I was born, and many others have essentially disappeared. So many glacial landscapes gone forever, so many photographs that can no longer be taken.

When you think of the North Pole, you probably think of polar bears and Santa Claus, and indeed, one of them definitely lives there. What you may not think of, however, is the shell of ice that floats upon the Arctic Sea. Different in nature from ice that builds up on land, the North Pole's sea ice naturally dwindles throughout the sunlit summer and then refreezes, incorporating recent snowfall during the cold, dark winter. For thousands of years, this pattern of losing ice during the warm season and adding ice during the cold season was enough to keep the North Pole on balance. But during the last half century, all that has changed.

The Arctic melting season still begins near the first of June, but the date of refreezing keeps getting pushed back as temperatures rise. In the 1970s, the melting season ended in September; nowadays it doesn't end until October. That extra month of melting translates into one fewer month of accumulation. The sea ice that covers the Arctic Ocean has thinned dramatically and its edges are breaking up, which is very bad news for polar bears, who need, among other things, somewhere to stand.

Most of the earth's ice is located at the North and South Poles, where the long, dark winters preserve and petrify each year's snowfall into the preexisting hard, cold ice. Thanks to the poles, about 10 percent of the earth's surface appears snowy white from space. Nonetheless, as the earth's overall warming accelerates, our satellites will observe those white patches shrink and eventually disappear.

Ice melts when the temperature adjacent to it exceeds thirty-two degrees Fahrenheit. This was probably the first science experiment you ever performed, when, as an infant, you reached into your mother's water glass, curious about the shiny cubes. She probably scooped out a few and let you hold them in your tiny hand, and you were enchanted, both by the glassy solid and by the water it left behind while melting.

I lost Covington soon after spring arrived in 1976. I've often thought that the image of six-year-old me discovering him reduced to a puddle one April morning, my tears mixing with his transubstantiated remains, would make a dynamite illustration for a children's book. It didn't go down like that, however. Winter stormed off and the world warmed up as usual; all the ice melted, and by May Day I had a real-live best friend named Jennifer.

To be honest, I don't even remember feeling sad about losing Covington. Perhaps I instinctively knew, long before I studied the hydrologic cycle, that he had gone where all melted ice eventually goes: to be enfolded into the wide, welcoming arms of the open ocean.

17

Rising Waters

Admit that the waters
Around you have grown.
                    —Bob Dylan (1964)

If there is something in the world nobler than a good dog, I have
not yet encountered it. All breeds have their charms, but I am
partial to the Chesapeake Bay Retriever—big, brown, and inde-
structible. I have owned several over the last thirty years and have
had the great good fortune of seeing all of them stubbornly per-
sist into old age.

We put Coco, our most recent Chesapeake, to sleep after
her kidneys succumbed to the wear and tear of a long life. My
husband and son and I sat in a little circle in the vet's office one
Tuesday morning. I held Coco in my lap and told her what a
good dog she was and how much we loved her. When her breathing slowed, then stopped, and I felt her heartbeat falter against

my own, I couldn't help thinking that this was not how I had expected her to die.

Chesapeake Bay Retrievers love to swim—they were bred to fetch ducks from the rough waters of the Atlantic Ocean—and Coco was an extreme example. While we lived in Hawaii, I always carried a coconut with me, for Coco often got it into her head to try to retrieve one of the great red buoys that floated offshore. Once she decided to swim out, a mighty splash from a fair-sized consolation prize was required to lure her off her course and back into retrieval mode, and a big rotten coconut always seemed to do the trick.

She was worst at Kailua Beach on the windward side of the island, where she would occasionally stop and stare out past the surf and across the vacant ocean, utterly bewitched. Thirty seconds later, she'd be off, struggling past the seven-foot waves as if her heart would break, riding the swells and swimming urgently toward California. I once swam out in a panic after the coconut didn't work, chasing her with my best front crawl and hoping that the undertow wouldn't kill us both before I could grab her head and turn her back to shore. Exhausted and shaken, we staggered back up the beach that day, both of us sputtering seawater, and I admit that I loved her all the more for her breakneck bravado.

Many more years went by and a frayed old coconut, combined with her love for the woman who threw it, was always enough to tempt her back to land. She is gone now, but I still have dreams where I am standing alone on the shore, watching her swim farther and farther out, until she is a tiny speck that disappears over the horizon.

After talking to my colleagues and looking at maps, I've come to believe that Coco was trying to retrieve the Waverider buoy that bobbed far out from the seaward edge of Kailua Bay, anchored at depth. I couldn't see the yoga-ball-sized metal sphere with my feeble human senses, but I'm convinced that Coco could smell it or hear the waves lap against it and somehow knew it was there. She was right to covet this marvelous instrument. Equipped with accelerometer, radar, and a Global Positioning System, the Waverider is a high-tech Chesapeake magnet built to relay precise measurements of heave, pitch, and roll back to shore nearly continuously and provide valuable information for surfers and sailors alike.

Hundreds of similar high-tech sensors polka-dot the oceans of the world, where they patiently track a wide range of ocean properties, including "vertical displacement," which means that they measure the exact distance from where they float down to the center of the earth, many times each day. A great army of satellites also watches the ocean's surface continuously, eager to record the slightest change in sea level. These instruments have greatly augmented the sheltered "tide houses" of old, which were basically floating sheds, often positioned at the end of a pier, in which a stationary ruler attached to a strip chart recorded the height of the surrounding water.

More than a century of such measurements, evolving as they did from the ancient art of tide prediction, shows that since 1880, the average height of global sea level has gone up by more than seven inches. More than half of this rise has occurred since I was born in 1969, which means that global sea level has not only risen, the process of rise is accelerating.

Though the thousands of simple instruments floating in the ocean clearly show this long-term global sea-level rise, it has not been the homogenous process that we might have expected it to be. Because both the ocean and the land are dynamic—waves swell and crash, while the land lifts and sinks—sea-level rise has been much higher than average in certain places, such as along the Texas Gulf of the United States, where the ocean has risen more than eight inches just since 1960. Here in Oslo, the land is rising faster than the adjacent ocean; it is still rebounding from the lost weight of melted glaciers, so the net level of Oslo Fjord has actually gone down by four inches since 1960.

About half of the global sea-level rise that has occurred during the last fifty years is due to water added to the ocean from melting glaciers. The other half of the sea-level rise is due to the warming of ocean surface waters. The oceans have absorbed most of the heat trapped by greenhouse warming, and seawater expands as it heats up. The average temperature of the ocean's surface waters has warmed by more than one degree Fahrenheit over the last fifty years; more than three inches of the total global sea level rise came just from the swelling of a warmer ocean.

The ocean's creatures have been thrown for a loop by this warming of seawater: off the coast of North America, the average fish has moved forty miles north and forty feet deeper in search of cooler waters. Of course, this has also changed the business of fishing: the average lobster has moved a hundred miles north since 1970, and many lobster boats have had no choice but to follow.

An increase in sea level has ramifications for life on land as well, some of them obvious and some not. When sea level rises, land is lost to the ocean: dry land—upon which we might build cities, hotels, homes, factories, or warehouses—is inundated and then disappears underwater. Just since 1996, the East Coast of the United States has lost twenty square miles of land to the open ocean—a net loss of valuable beachfront property. Other effects of rising seas take more time to show themselves: the flooding of farmland and groundwater with salt water forever ruins the fertility of the soil and the drinkability of the water.

At present, one-quarter of the earth's population lives less than sixty miles from the coastal ocean. Even with the best protection that engineering can provide—a dike system similar to that undertaken in the Netherlands, perhaps—sea-level rise will likely displace thousands of people during the next hundred years and make drinking the local water and farming the adjacent fields impossible for many thousands more. These effects will hit worst in areas where the land is low or sinking and where people are too poor to undertake construction measures.

The river-delta nation of Bangladesh lies just barely above sea level. Within its borders, a population half the size of the United States ekes out a living from a piece of land the size of Alabama. If the sea continues to rise, the area of Bangladesh is likely to shrink by 20 percent over the next thirty years, crowding people into even less land and fewer resources. Incidentally, the people of Bangladesh produced far less than 1 percent of the carbon dioxide emitted to the atmosphere during the last fifty years, yet they are poised to pay the highest price incurred by its effects. This is a common trend: the people benefiting from the use of fossil fuels are not the people who suffer the most from its excess.

Carbon dioxide has not only affected the ocean indirectly through warming, it has altered it directly by soaking into seawater. Almost one-third of the carbon dioxide produced from the burning of fossil fuels was absorbed by the ocean, like the way carbon dioxide gas is injected into Coke (with chow; wow!) in order to make it fizzy. Your dentist has probably warned you that drinking soda is bad for your teeth, and part of the reason is that carbon dioxide reacts with water to form an acid which eats away at your enamel. This same thing is happening in the ocean, as the increased acidity of seawater has decimated coral reefs around the world, making it harder and harder for shelled marine animals to grow and maintain their hard parts. As we go about our business burning fossil fuels, more carbon dioxide enters the atmosphere, and thus the ocean, every day.

Back to melting for a moment: Remember how half of the sea-level rise that has occurred during the last fifty years is due to water being added from melting ice? The ice that has already melted and entered the ocean, causing a rise in sea level that displaced thousands of people, amounts to less than 5 percent of the ice that *could* be melted by future warming.

All of the ice that I mentioned in chapter 16 (namely, the world's glaciers and the sea ice covering the Arctic Ocean) is dwarfed by a comparison to the vast ice fields that blanket Greenland and the continent of Antarctica. The wide expanse of ice that covers the rocky surface of Greenland has begun to falter: like the rest of the Arctic, it is melting, losing billions of tons of ice each year. Fortunately, for now, some of this meltwater refreezes as it runs across the surface of the ice sheet, but some does enter the ocean.

And then there's the Antarctic ice sheet, fields of ice that are more than a mile deep and a million years old, on which stand a full 90 percent of the ice on planet Earth. So far we have seen less melting in Antarctica compared with the Arctic, but this could change: radar sounding waves have shown that the ice is beginning to peel away from the portion of Antarctica that faces the broad expanse of the Pacific Ocean. Should continued warming melt the Antarctic ice sheet in earnest, the amount of water released into the ocean, along with the subsequent sea-level rise, will be massive.

The global records of carbon dioxide, temperature, ice mass, and sea level are large data sets from simple measurements that show an obvious trend over the last twenty years and yet somehow, they also give rise to nonstop drama. I can't turn on my computer without hearing about climate deniers and their lack of brains and sophistication. I can't click on any of the links within without hearing even more about climate alarmists and their hypocrisy and hyperbole.

Separated into two camps, we flame each other across the Internet as if the atmosphere cared what we think, as if we could shout the rising waters back up into the glaciers they came from, as if winning an argument accomplishes something in and of itself. America has become that unhappy couple who is mired in fights about dishes and laundry because both partners are too terrified to earnestly examine any sort of change.

In the meantime, scientists continue making the observations and taking the measurements that show what is happening. I won't lie to you—it's a rough gig. Today's climate scientist is more like a worried oncologist than like an Indiana Jones or a Jacques Cousteau. The earth is sick and we suspect that it's something bad. This heat, this melting, this flooding, these hurricanes, those blizzards—they are not normal, and we need to do the tests that allow for a firm diagnosis and a treatment plan. We hope that it is not as bad as it looks, but we need some support, cooperation, and a little patience to get to a definitive answer.

The carbon dioxide level of the atmosphere is higher than

it has been at any time during the last million years. The ice is melting, the sea is rising, and the weather has begun to falter. Our earth is unwell. Its symptoms are worrying and will not improve on their own. We have all but used up the brief window when we can act, and we are running out of time. I know this because many of the things around us have already started to die.

18

The Big Good-bye

*Great blunders are often made, like large
ropes, of a multitude of fibers.*
— Victor Hugo (1862)

I was fully prepared to encounter plenty of dead fish when I
opened the door of the PUCRS Ichthyology Labs in 2005. What
I wasn't prepared for was the smell. It was enough to physically
knock you back. It was enough to make your eyes water. It was
enough to make it hard to hear the person talking to you. It
was impossible to focus on anything but the overpowering smell
of booze. That lab smelled like the inside of a vodka bottle—if
that bottle had been immersed in a bathtub of whiskey and sunk
to the bottom of an ocean of gin.

The Pontifícia Universidade Católica do Rio Grande do
Sul (PUCRS) is located in Porto Alegre, in southern Brazil, just
near the mouth of the Jacuí River. It contains a gleaming state-

of-the-art dissection laboratory that has been assigned the Herculean task of compiling an up-to-date list of the freshwater fish of South America. It is one of many institutes created for the sake of documenting the incredible diversity of flora and fauna that exists within the borders of Brazil, a land so lousy with butterflies, insects, flowers, fish, and amphibians that it has stretched the imagination of modern biologists completely out of shape.

As of today, thousands of species of freshwater tropical fish are known to exist, and about half of them live in Brazil. Of those living in Brazil, the great majority is also *endemic* to Brazil—which means they are native species living in their original habitat, a rare thing in this age of global disturbance. The PUCRS lab that I visited was chin-deep into its decades-long collection project: teams of students and postdocs had set nets across the rivers of southern and eastern Brazil. From each of these rivers they hauled hundreds of fish back to the lab, sorted them into piles, and preserved them in grain alcohol until they could fully examine and describe them all. The sheer variety of what they brought home in those nets was brain boggling. For most scientists, identifying a new species is a career-changing, once-in-a-lifetime event; these South American biologists were discovering a new species, on average, once every four days.

Only after amassing a detailed anatomical profile of the fish would they be able to add it to the official count. They would verify whether its species had been previously seen; if it was a new species, they would determine to which others it was most closely related. It was downright comical how many fish were stored in that lab, all of them immersed in ethanol and waiting their turn. Against every wall—of the lab, of the hallway, on the balcony—stood a line of white plastic barrels, tall as my shoulders. A single handle was attached to the top of each lid, and when you lifted it, up came forty or more fishing lines, each one with a dead fish at the end. I would have been only mildly surprised to see a fish floating in the coffee cup that my tour guide was drinking from.

"Will you ever get to all these?" I asked her as she began the paperwork for assimilating the latest haul.

"Eventually, yes," she assured me, sipping from her cup. "We already see them tapering off quite rapidly, in fact." She added, by way of explanation, "A lot of what we do here is about looking a fish in the face at least one time before it goes extinct."

There's a reason that the PUCRS lab is putting so many resources into cataloging the freshwater fish of Brazil: they are the canary species in the ecological coal mine. Freshwater habitats are exclusive: they make up less than 1 percent of Earth's surface but contain at least 6 percent of our planet's species. This high diversity makes any change in population all the more dramatic. The number of endangered species within freshwater habitats is greater, proportionally, than in the land and ocean combined.

The Jacuí River, which flows past the PUCRS lab and joins with Lake Guaíba on its way to the Atlantic Ocean, has a story that's pretty typical for a river in Brazil that is located near a populated area. In 2000, the upper waters of the Jacuí were stopped up when the Dona Francisca Dam was built. On the upriver side of the dam, five thousand acres of land were flooded to create the Dona Francisca reservoir, and more than five hundred families had to relocate their homes as a result.

Dams are built for all kinds of reasons, but they all result in the same thing: an upstream lake where there previously was none. Lake water is more accessible than flowing water because it can be withdrawn in volume to irrigate crops, water animals, and wash or cool machinery inside factories. Dammed-up lake water can also be released over the dam in order to generate electricity, as described in chapter 13. Dams are built to support industry and agriculture, and by doing so, they almost always give rise to more industry and more agriculture.

When a dam is built, life downriver is forever altered. The most obvious change is that the volume of flowing water is reduced to a small fraction of what it was when the river was running freely. Once a river's flow is reduced, its real estate available to fish, insects, and amphibians is also reduced. Crowding ensues, competition increases, resources become strained, and many of the river's inhabitants die. Hence the race against time within the biology labs of Brazil: they want to figure out which fish are there today, so that we will know which ones have disappeared tomorrow.

Extinctions are part of the natural course of action on planet Earth; a new food source or living space opens up, and this creates a niche. A subset of one species hones its skills for life in the niche, they move in and breed amongst themselves, and before long, a new species is born. Eventually, the niche closes after the environment changes, or a stronger competitor moves in. Once the niche is gone, its specialized species declines and eventually goes extinct. Paleontologists see this process over and over in rocks containing fossils: species are constantly popping up and constantly disappearing. The average species persists for something like ten million years. Because this cycle of emergence/extinction has been in place for more than three billion years, vastly more species have been lost to extinction than exist today.

Very infrequently, a whole bunch of niches will close at the same time, triggering what is known as a mass extinction. The fossils of the last five hundred million years show evidence for five such biological catastrophes: short periods during which 70 percent (or more) of existing species were lost to extinction, all at once. Earth's most recent mass extinction occurred about sixty-six million years ago and resulted in the disappearance of the dinosaurs. Based on the rate of decline and disappearance of species today we fear that we are currently on the verge of a sixth mass extinction.

The large-scale species surveys of the last few decades show a disturbing trend: a decline in more than half of all bird and butterfly species and a complementary decline in one-quarter of all fish and plant species. We are witnessing rapid losses before our very eyes, from the base of the food chain on up. For several species, we have seen wild populations decline into extinction: the thicktail chub, the Rocky Mountain locust, the Japanese sea lion, the black softshell turtle, and the stringwood tree, to name a handful.

The first reason for today's decline in species is the simple, and widespread, loss of habitat; expanded cities mean fewer places for plants and animals to live. The history of North and South America shows this most plainly, as European settlement progressed into full-scale colonization, all within the last five hundred years. Since the first Europeans showed up, North and South America have lost 88 percent of their tropical forests, 90 percent of their coral reefs, and 95 percent of their tall-grass prairies. The species that were adapted to those niches have all but vanished now. The cities, suburbs, harbors, and farm fields that we built presented brand new niches, ones just right for the "invasive" microbes, insects, plants, and animals that came with the Europeans, saw, and conquered.

The second reason for modern species decline are the climate changes we covered in chapter 15. Extreme heat waves have been known to kill whole colonies of bats, and disappearing polar ice has decimated the hunting ground of polar bears. Elevated levels of carbon dioxide fertilize the growth of poison ivy over that of native trees, and the future of green sea turtles has been doomed by the warming that converts their eggs into exclusively female spawn. In many cases, previously natural areas have been made over into more uniform and boring lands of concrete, asphalt, ditch, yard, and hedge. Within them, just a few non-native species flourish at the expense of the many and varied original tenants.

The current rate of global species extinction is almost one thousand times higher than the background rate seen in the rock record of fossils. If we project from today's rate of loss, the year 2050 will put us close to a total of 25 percent extinction of species; that's a third of the way to the 70 percent mark that characterizes a mass extinction.

Geologists have written countless volumes about the ancient rock layers encompassing mass extinctions. The layers from the time before, after, and during the five mass extinctions that occurred millions of years ago have been surveyed and sampled, analyzed and archived. This sleuthing revealed some kind of environmental catastrophe associated with each mass extinction: an asteroid impact, abrupt sea-level change, sudden greenhouse gas release, gushing lava flows, unchecked flooding or fire—the list goes on. Cataloging the fossils lost during a mass extinction never fails to upend the earth's genealogy, for when a species is lost to extinction, its descendants are lost also.

One important thing that separates a mass extinction from a more run-of-the-mill extinction is the rapid pace of species loss; in a mass extinction, the total species loss occurs over thousands of years instead of the millions that pass while background extinctions come and go. The mass extinctions of long ago were dramatic periods of destabilization that resulted in widespread reorganization—exactly what we are starting to experience today.

At the end of a mass extinction, the tree of life has lost several branches—and yet, afterward, life does go on. Plants regreen the earth and animals repopulate the oceans; different species take over and different landscapes result; and time resumes its relentless forward march. There will be life on planet Earth after the sixth mass extinction, but we are not able to imagine it any better than the dinosaurs could have imagined a world dominated by mammals walking on two legs, driving bulldozers, and flying airplanes.

All species will go extinct eventually, even our own: it is one of nature's few imperatives. As of today, however, that train has not quite left the station. We still have some control over our demise—namely, how long it will take and how much our children and grandchildren will suffer. If we want to take action, we should get started while it still matters what we do.

19

Another Page

To love, and bear; to hope till Hope creates
From its own wreck the thing it contemplates.
—Percy Bysshe Shelley (1818)

When I was a little girl, my father told me that one day, all the people on Earth will live together, free from war and hunger and need. When I asked him when it would happen, he said he wasn't sure but that he was sure the day would come.

These many years later, I wonder if he chose that day, in particular, to talk to me. I had just turned nine, which made me old enough to question his views but still young enough to listen to his answers.

"How?" I asked him; I was skeptical. "We all live in different countries," I observed, "and we all speak different languages."

"Borders can change," he responded quietly, "and we can

learn each other's languages." He added, "We are a species that can learn anything."

I was his youngest child, his only daughter, and I had come to him late in his life. He was fifty-five years old when we had this conversation and had already seen the world map revised twice during his lifetime. But this was not the only reason he believed in our ability to transform the world.

He believed it because he had seen people grow and change when he taught physics, chemistry, calculus, and geology to an entire generation of our town and then to the entire generation of their children. He believed it because of what he had seen during his own life: the magic of radio had become television, the telegraph became the telephone, the computer with ticker tape became the computer with punch cards, which eventually became the magic of the Internet. He believed it because of his family: his children had a mother who had survived their birth (unlike his grandmother), would have the opportunity to go to college (unlike his parents), and were growing up free from the shadow of polio (unlike him).

He believed it because he loved me, and because of him, I believe it too. I believed my first, and my fondest, science teacher when he told me that when we finally devote ourselves to work and to love, even our most fantastic dreams will eventually come true.

Multiple solutions have been proposed to stabilize, and then decrease, the amount of carbon dioxide that is in the atmosphere in the hopes of limiting—and ultimately reversing—the earth's rising temperatures, and all of these solutions show some promise. It is possible to separate carbon dioxide from air, concentrate it, and then seal it into a container; the torpedo-shaped metal tanks standing next to a restaurant's soda dispenser are an example of this technology. Engineers have proposed to remove carbon dioxide from the atmosphere, condense it into a pure liquid, and then inject the liquid deep into the earth, where it would remain forever trapped between layers of rock at the bottom of the ocean or in the caverns left behind after drilling for oil and mining for coal. The unfortunate problem, so far, is that this process of trapping, condensing, transporting, and then injecting takes more energy than it compensates—that is, the amount of fossil fuel that you have to burn in order to make it happen is more than the amount of burned fossil fuel you are able to trap in the end. Nevertheless, several nations continue to refine these "carbon capture and storage" methods into something that might someday break even.

Equally fanciful is the notion of drawing down carbon dioxide by enhancing rock weathering: every year, a small amount of carbon dioxide naturally leaves the atmosphere to mix with rainwater. This yields a weak acid that trickles down into the soil and dissolves the bedrock at its base. Scientists have proposed to increase the amount of rock surface-area available for weathering by grinding up volcanic minerals and then dusting the debris across the tropical forests and farm fields of India, Brazil, and Southeast Asia. As with carbon capture and storage, however, the energy used in grinding, transporting, and spraying millions of tons of rock dust would outbalance any gains that were made in carbon dioxide reduction, at least for several centuries.

One popular solution is to stimulate plant growth, to naturally pull carbon dioxide out of the atmosphere and transform it into living tissue. This is what trees do by growing on land, and it is what algae do by growing in the ocean. Multiple schemes offer people an offset of the carbon dioxide produced when they fly by paying money for trees to be planted. This is an appealing concept—after all, who doesn't like the idea of more forest?—but the numbers are not as favorable as you might think. First of all, the rewards are slow to kick in, as most seedlings take decades to grow into a respectable tree. Furthermore, trees don't permanently incorporate much carbon dioxide into their tissues, as most of the leaves and needles that fall to the ground each year shrivel up and decay, releasing carbon dioxide right back up into the atmosphere. The most stable stash of fixed carbon dioxide in the forest is the tiny percentage of rotted plant material that makes it into the soil, and new soil takes hundreds to thousands of years to form. Planting trees will offset your energy use, but it won't pay off during your lifetime and possibly not during your children's lifetimes, either.

Moreover, even if we could reverse all the deforestation that has taken place since 1750, replanting all the forests that have been cut down, the carbon dioxide stored in these newly planted forests would equal, at most, about two decades' worth of carbon dioxide emissions at today's rate of fossil fuel use. We'd also be faced with an urgent dilemma—namely, we'd have no place to grow crops, for most of the deforestation that has occurred has been for the purpose of creating farmland.

More promising is the idea of fertilizing plant growth on the ocean's surface: there are vast areas of the Pacific Ocean that lack only one or two key nutrients and could be stimulated to yield huge colonies of green algae and phytoplankton within weeks—tiny plants that would eventually die and sink to the bottom of the ocean, effectively trapping the carbon dioxide that they used to form their tissues. Seeding tropical waters with iron or phosphorus could trigger such blooms, but the exact amount of carbon dioxide that could be trapped is uncertain, as are the consequences for ocean life, including the fish that we depend upon as a food source.

An alternative approach for reducing the earth's temperature is to reduce the amount of incoming sunlight—to simply block the sun's energy before it interacts with Earth's atmosphere. Ideally, this approach would allow for cooling without requiring any changes in fossil fuel use. Many ways of blocking the sun's rays have been proposed, including launching a huge sunshade into orbit in order to reflect heat outwards, similar to the one you place under your car's windshield on hot days. Another idea is to spray aerosol particles high above the clouds, in much the same way that Mount Pinatubo did during 1991 when it erupted, leaving a layer of haze that cooled portions of the planet by almost one degree Fahrenheit for two full years. These approaches have the advantage of being quick and resulting in substantial cooling almost immediately. They are, however, incredibly risky: tinkering with the heat balance of the atmosphere is almost certain to disrupt weather, and less precipitation and more frequent droughts are the last things we need for global agriculture.

As for dealing with sea-level rise, there is already a long tradition of successfully engineering cities such as Amsterdam to resist flooding, but the techniques are expensive and out of the reach of many poor countries, such as Bangladesh. Novel constructions have been proposed to directly mitigate melting polar land ice, which will likely be the main source of added ocean water during the next several centuries. These constructions include a concrete wall in western Greenland to prevent warm seawater from coming into contact with the glaciers' base.

Even more ambitious is the proposed plan to prop up the icy edges of West Antarctica from underneath via an artificial island anchored to the seafloor. The idea is that such a berm would serve to keep the ice sheet whole, and prevent it from fragmenting and then melting.

Moving tons of concrete and gravel and using heavy machinery in polar regions seems outrageously expensive, but perhaps less so when we consider that the cost of the world's largest construction projects, such as the Hong Kong International Airport or the hydroelectric Three Gorges Dam, commonly run to $20 billion and $30 billion.

Regarding the crisis in species decline, nothing has been shown to prevent extinctions so well as simply protecting habitats by preventing human access and development. About 13 percent of Earth's land area is currently under some degree of legal protection, which is more than triple the amount that was protected forty years ago. Ecologists believe that this increase may have reduced extinction rates in mammals, birds, and amphibians by as much as 20 percent, relative to what they would be otherwise. Multiple nations are also considering the establishment of marine protected areas, conserving swaths of ocean from both fishing and shipping transport.

The "father of biodiversity," E. O. Wilson, is now promoting the ideal of the "half earth"—a full 50 percent of the earth's land to be designated a human-free natural reserve. We should never underestimate the potential for private individuals to alter the trajectory of global change: In 1977, United States poet laureate W. S. Merwin started planting trees on a dumping ground in Maui. Four decades later, those nineteen acres contain more than four hundred species of tropical trees and preserve the most critically endangered palm trees in the world. These are the types of measures needed if we hope to prevent a sixth mass extinction during the next several centuries.

None of the solutions above, however, address the root of the problem or have the potential to alter our future the way a serious shift toward energy conservation could. I mentioned back in chapter 10 that if all the fuel and electricity in use today were redistributed equally across the earth's population, global per capita energy use would be equal to the average consumed in Switzerland during the 1960s. If, instead of waiting for this imaginary redistribution, the countries of North America, Europe, Japan, Australia, and New Zealand reduced their energy use to that level now, total global energy use would plummet by at least 20 percent, as would carbon dioxide emissions.

The changes to daily life required to meet this reduction would be most keenly felt in the United States, where energy use per capita is the highest in the world. Each American would need to forgo four out of every five plane rides and travel using mass transportation at least fifty times farther each year than they do now. The United States as a whole would have to get rid of at least 30 percent of its motor vehicles, and the resultant effect on the trucking industry would require that we eat and shop from an entirely different set of goods.

The good news is that there is no reason to think that energy conservation will necessarily reduce our quality of life: life expectancy in Switzerland back in 1965 was similar to that of today's America and much higher than the current global average. Workdays were shorter, as were commuting distances. Life was not perfect then, but the basics of a healthy life were already in place at a much lower level of fossil fuel use.

Indeed, if we look to the most comprehensive measures used to estimate the elusive concept of "happiness," we find that our increasing consumption of food and fuel over the last decade has not made us happier—quite the opposite. In 2017, the Global Happiness Council, a group headed by United Nations adviser Jeffrey Sachs, reported that Americans were the unhappiest they had ever been, at least since 2005, despite the fact that they were working, eating, driving, and consuming more than ever before.

According to that same report, analyses from more than 150 different countries determined six factors that form the social foundations of the cross-cultural concept of happiness: social support, freedom to make life choices, generosity, absence of corruption in government, healthy life expectancy, and per capita income. It goes without saying that most of these factors can be maintained, or even improved, while reducing fossil fuel use.

As a solution, energy conservation by its very definition requires the least effort of any approach. It is a strong lever by which we could pull ourselves back into alignment with a future that our grandchildren might survive. There's only one problem: driving less, eating less, buying less, making less, and doing less will not create new wealth. Consuming less is not a new technology that can be sold or a new product that can be marketed, and acting as if it can be is absurd. Economic proposals such as the "carbon tax" and other good-faith attempts to provide monetary incentives against fossil fuel use have been met with terminal opposition from industrial sectors.

It's no use pretending that conserving resources isn't at direct odds with the industries that helped to write our Story of More and that increasing consumption over the last fifty years wasn't tightly coupled to the pursuit of *more* profit, *more* income, *more* wealth. It's time to look around and ask ourselves if this coupling is truly the only way to build a civilization, because the assumption that it is may represent the greatest threat of all. Each one of us must privately ask ourselves when and where we can consume *less* instead of *more*, for it is unlikely that business and industry will ever ask on our behalf.

None of the solutions discussed here, alone, will be a silver bullet in terms of saving us from the dark specter of hunger, want, and suffering that unabated consumption is likely to bring about. All measures of conservation, as well as all technologies meant to wean us from fossil fuels, are worth pursuing, in the same way that doing something is always more than doing nothing. We need everyone, not just scientists, to start thinking about tomorrow. We need to figure out not only the possibilities associated with each solution but also the risks, so that whenever we have the chance to act, we can do so with our eyes open and our understanding as full as we can make it.

Earth—the only thing we all share—has become a pawn in our political discourse, and climate change is now a weapon that can be hurled by either side. For scientists especially, feeding into political pique and polarization actively damages the planet we are trying to save. Down the road, it won't matter what we do as much as it matters what we *all* do. Need I mention that "we all" has always included—and always will include—both me and you? We are all part of what is happening to the world, regardless of how we feel about it, regardless of whether we personally "believe" or "deny." Even if you consider yourself on the right side of environmental issues and a true believer in climate change, chances are that you are actively degrading the earth as much as, or more than, the people you argue with. An effort tempered by humility will go much further than one armored with righteousness.

During each of the years that I've taught this material, at least one student—overwhelmed by the data—would come to my office and ask me if I thought there was any hope for planet Earth. Here is how I would answer:

Yes. Absolutely. I do believe that there is hope for us, and you are very welcome to take some and keep it for your own.

I am hopeful because my life is filled with people who also care about these issues. The smartest people I know are dedicating their lives to gathering the data that will tell us more. This very day, scores of people got to the lab early and will stay late, trying to quantify the exact magnitude of sea-level rise, warming weather, melting polar ice. They are walking the fields and counting what is there and what is not. The ecologists who first noticed these patterns could not have imagined the computers or instruments that we now use every day. We are watching and working and not just worrying. Climate science is a part of science, after all, and science is now just as it has always been: overworked and underfunded and absolutely unwavering in its refusal to ever stop trying to figure it all out.

I am hopeful because history teaches us that we are not alone. During centuries past, women and men railed helplessly against overwhelming forces that poisoned the wells, spoiled the crops, and robbed them of their loved ones. We may discount their science as superstition, but it was based on state-of-the-art observations and earnest conclusions. Genetically we are no smarter than they were, and we may be laboring in similar darkness. The succeeding centuries did bring unfathomable solutions to even the most intransigent of these ancient plagues, and though the solutions came far too late for many, they were not too late for all.

Then comes the hardest part of the conversation, when I ask them to think about their own lives.

I remind them: We are strong and lucky. Our planet is home to many who struggle to survive on too little. The fact that we are of the group with food, shelter, and clean water obligates us not to give up on the world that we have compromised. Knowledge is responsibility.

I ask them: What will you do with the extra decade of life given to you, over and above your parents'? We, the 20 percent of the globe that uses most of its resources, must begin to detox from this consumption, or things will never get better. Look at your own life: Can you identify the most energy-intensive thing that you do? Are you willing to change? We will never change our institutions if we cannot change ourselves.

I stress one thing above all else: *Having hope requires courage.* It matters not only what we do about global change but how we talk about it, both in the classroom and beyond. We risk our own paralysis with the message that we have poisoned the earth and so the earth rejects us. As far as we know, this is still our species' eternal home, and we must not alienate our children from it. We must go forward and live within the world that we have made, while understanding that its current state arises from a relentless Story of More. We can make this easier by being kind to one another along the way.

I warn them: Do not be seduced by lazy nihilism. It is precisely because no single solution will save us that everything we do matters. Every meal we eat, every mile we travel, and every dollar we spend presents us with a choice between using more energy than we did last time or less. You have power. How will you use it?

Now is the time to imagine a world in alignment with our ideals, as we embark on our postindustrial age. We will need to feed and shelter ourselves and one another, but everything else is on the table. What can seven billion people do that three billion people could not? is the question of my life so far. We are troubled, we are imperfect, but we are many, and we are doomed only if we believe ourselves to be. Our history books contain so much—extravagance and deprivation, catastrophe and industry, triumph and defeat—but they don't yet include *us*. Out before us stretches a new century, and its story is still unwritten. As every author will tell you, there is nothing more thrilling, or as daunting, as the possibilities that burst from a blank page.

Appendix: The Story of Less

Our healing is not in the storm or in the whirlwind, it is not in monarchies, or aristocracies, or democracies, but will be revealed by the still small voice that speaks to the conscience and the heart, prompting us to a wider and wiser humanity.

—James Russell Lowell (1884)

I. The Action You Take

It is in you that you must look, not around you.
—Eugène Delacroix (b. 1798, d. 1863)

Now, after all you've read, I have one question for you: Do you want to live in a more equitable world with a brighter future?

If your answer is "Yes," then we need to talk about the steps necessary to get there, bearing in mind that while Rome wasn't built in a day, it didn't burn down in a day, either. Take some time to think about your answer, and talk to those closest to you, for this will help make your actions stick.

Step 1: *Examine your values.*

A large number of issues have been raised in these nineteen chapters. Which ones resonate with your daily life, your greatest fears, your highest aspirations? Consider them all, then rank them accordingly. Where does world hunger fall on your list? Extinction of species? Weirding weather? Green energy? Pollution of our oceans? Animal rights? Public transportation? Beach erosion? Healthy school lunches? National parks? Organic farming? Arctic warming? Women's health? Some of these issues may be very important to you and others less compelling. Identify one issue to focus upon: the one that you are willing to sacrifice for.

Step 2: *Gather information.*

Go through your habits and possessions in order to take stock of how, as with most of us, your personal life is working against your values. How many miles do you drive? How often do you fly? Is any of it optional? How much of the water that goes down your drain is still drinkable? How much of the food that goes into your garbage is still edible? How much meat do you eat, and how often? Go through your closet and look at the tags: Where were your clothes made? How far did they travel to get to you? Go through your refrigerator: How many items came to you in a plastic container? How many of them contain added sugar, posing as "natural sweetener"; "corn syrup"; "cane sugar"; "cane syrup"; "maltodextrin"; "fruit juice concentrate"; "raw sugar"; "brown sugar"; "dextrose"; "glucose"; "HFCS"? Is nearby land being contested for development versus conservation? How is the electricity that comes into your house generated? Is constructing a "renewable energy" facility up for discussion within your county? Where does the gasoline you pump come from? Is any of it ethanol? Which of the meats that you consume concentrates the most grain? Which parts of your sushi were likely raised inside a farm?

Step 3: *Can you make your personal activities consistent with your values?*

Pick out one change that you can make. Can you drive fewer miles? Carpool? Eliminate some of your plane trips? Use public transportation? Buy 40 percent less food (especially the items you notice finding their way to the garbage)? Avoid sugared foods? Eat fewer meals of meat each week? Reuse plastic objects more than twice? More than three times? Keep your thermostat lower during the winter and higher during the summer? Buy locally? Buy *less*? Give up *more*?

Be sure to keep a journal on how it goes; record the numbers and the outcomes. After you take these first three steps, you'll find yourself knowledgeable, experienced, humbled, and proud with respect to your values—all of these are necessary, though not sufficient, qualities for convincing others.

If you've made it this far, please know that I am very proud of you for many reasons, not the least of which is that now you are ready for the hard part.

When I was in high school, one Minnesota physician caused a scandal that reached even our small-town newspaper. A renowned transplant doctor—a surgeon who could swap out your liver or kidneys—was exposed as the owner of a Popeyes Famous Fried Chicken & Biscuits outlet.

"He gets you coming, and he gets you going," my mother wryly observed before turning the page.

To the reporter, the doctor had defended his investment on health grounds. "The important thing is to get away from beef and red meat," he lectured. "What we stress, as physicians, is chicken and fish." I had never been to a Popeyes (there wasn't one in our little town), but the journalist had also contacted a Harvard professor who assured readers that Popeyes food contained just as much saturated fat as the burgers from any of the big burger joints. Probably bad for your kidneys, he surmised, and definitely bad for your liver.

Hypocrisy and greed are not personal qualities that midwesterners admire, though the surgeon continued to practice. He may have been pilloried in the *Star Tribune,* but it's quite likely nobody ever said anything to his face. As of today, Popeyes is still in business and making more money than ever.

At the time of the scandal, I was staring down the barrel of four long years at university. As a surgeon, this doctor had studied for at least eight years, followed by several more years of twelve-hour days fading into countless night shifts as an intern and then a resident. If he had been motivated solely by a lust for money, I wondered, couldn't he have found some easier way to earn it that didn't involve twenty-odd years of torturous training?

What I really couldn't fathom was why he would invest in something that worked so directly *against* what he was trying to do every day at his job. Did he want people to have healthy livers or not? The hours of his life pledged yes, while his acquired possessions screamed no.

Twenty years later, I found myself in almost the exact same situation.

In 2008, my lab published our chemical analyses of fast-food meals from McDonald's, Burger King, and Wendy's restaurants spanning the length of America, from Boston to Los Angeles. We were shocked to find that the meat in a sandwich purchased in Detroit had an identical isotope signature as the same sandwich purchased in Denver, or Cleveland, or San Francisco. Honestly, our results were consistent with the hypothesis that one single gargantuan chicken dwelled in captivity somewhere in Nebraska, and every time Wendy's sold a sandwich, a slice was taken from its massive breast.

Within our scientific publication, we attributed this extreme homogeneity to the industrially prescribed diet of farm animals and the confinement they endured before slaughter. Our paper didn't include the phrase "so stop eating that junk," but the overall picture it painted of the fast-food industry was less than savory.

We published our findings in a top journal, got some media attention, and I was feeling pretty good about myself. I personally hadn't eaten fast food since 2004, when, eight months pregnant, I demanded we go to a drive-through, wolfed down a double-cheesy-bacon whatever, and became overwhelmingly sick immediately afterward. My abstention of fast food since then had cleared my path to the pulpit, or so I thought.

Exactly one month after the 2004 Burger Incident, I was holding a beautiful baby and everyone around me was urging me to start saving—*investing*—against the End of the World, or baby's college, or whichever came first. We dutifully broke open our piggy bank and took the pennies out to buy stocks and bonds because, well, that's what one does. Four years later, upon publishing the study, I began to crow over my erudite takedown of Big Fast Food (is there any other kind?). My husband, Clint, looked puzzled and reminded me, "You know, we're probably invested in that."

Step 4: *Can you make your personal investments consistent with your values?*

The above events spurred a financial cleansing that continues within our family to this day. It takes time, as it's far from easy to figure out exactly what the hell you have invested in, thanks to index and mutual funds that effectively bundle stocks you would be proud to own with stocks that you would be mortified to be any part of. Picking individual stocks is also full of pitfalls. If you own "QSR," a nicely performing company bearing the nebulous moniker "Restaurant Brands International," you are actually invested in just three entities: Tim Hortons, Burger King, and—who knew?—Popeyes. If you own stock that finances activities that are directly orthogonal to what you are trying to accomplish with your life, consider pulling your money out. Are you an addiction counselor paying into a mutual fund that includes a tequila manufacturer? Are you advocating for low-income housing while paying into an index fund supporting the very company that's trying to gentrify?

Of course, buying stocks and bonds is not the only way to invest your assets; every time you make a purchase you are investing in something. When you buy a cappuccino on the run, you are investing in the location of the café, how it treats its employees, the means by which it obtains coffee beans, the living conditions of the sourced milk cows, and the transport system that brought all those ingredients into your neighborhood. It's admittedly a bewildering challenge, but it's useful to think about which of the five things on that list conforms to your values and which ones do not. Can you patronize a café that meets two of your criteria, while you search for one that meets three? Baby steps. No one can run before they can walk.

Step 5: *Can you move your institutions toward consistency with your values?*

At this point you possess the magic criterion to advocate for change: personal experience. Go to your children's school, your house of worship, your place of employment, and start a dialogue with those in charge. Share your values, your struggles, and your experience. Listen to them talk about their restrictions and concerns. Thank them for their time. Write a follow-up letter emphasizing your values, your struggles, and your experience, and ask for another appointment. Talk to your peers within the institution. Keep going back, keep advocating for the things that you believe in. It takes time and perseverance, but people (even politicians) and their institutions can change.

But you're only one person in a world of more than seven billion. Will changing your values make any difference?

II. The Difference You Make

*Give me a big enough lever, and a place to
stand upon, and I will move the Earth.*
　　　　　　　　—Archimedes (c. 250 B.C.)

Making a difference as to global change is often about finding the biggest lever, figuring out where to stand, and then pushing and pulling like hell.

For better or worse, as a member of the Organisation for Economic Co-operation and Development (OECD), you are part of the biggest lever in the world, with respect to the issues raised in this book. The people of the OECD, while representing only one-sixth of the global population, use one-third of the world's energy and half of the world's electricity and are responsible for one-third of the world's carbon dioxide emissions. Not only that, but the OECD consumes one-third of the world's meat as well as one-third of the world's sugar.

For even better or even worse, America is by far the biggest lever within the OECD. With a population amounting to one-quarter of the OECD (and only 4 percent of the entire world), the United States uses half of the OECD's total energy and one-third of the OECD's total electricity and is responsible for just under half of the OECD's total carbon dioxide emissions. The United States also eats one-third of the OECD's total meat as well as one-quarter of the OECD's total sugar.

This means that if you are a citizen of the United States or any of the other OECD countries, any steps you take toward austerity will have an outsize effect on global consumption. Let me give you an example:

Let's say an examination of your values has led you to the belief that reducing carbon dioxide emissions is important. A fact-finding mission revealed that all of your household electricity comes from a coal-burning plant on the other side of town. Because at least 20 percent of the total energy we consume comes to us as electricity, you've decided to take action by cutting down on your home usage of electricity as much as possible. How to proceed?

A good way to start is to discern the biggest lever in your home, in terms of electricity usage. In the European Union (EU), all new appliances are required to be sold bearing an "ENERG" sticker displaying several indices of energy efficiency. The critical value to look for is the number of "kWh/annum," which is the estimated amount of electricity that the appliance will use over the course of one year while serving a household of five. By comparing this number between makes and models, members of EU countries can begin to choose and control the level of energy use in their homes. Some American companies (General Electric, for example) provide kWh/year estimates for selected products, such as freezers and refrigerators; additional companies could be lobbied to make this information more available.

For most homes and apartments around the world, it is by far the electric water heater that uses the most energy; having hot water usually amounts to about half of the total electricity used in your home. If you're able to change from a fifty-gallon water heater to a twenty-gallon model and use cooler water (or even cold water) while washing clothes, rinsing dishes, and taking showers, you can cut your water heater's electricity use by as much as half, meaning that you've now cut your total household electricity use by one-quarter.

The next biggest levers are the machines designed to heat large spaces and to cool them—that is, space heaters and air-conditioning units. Together, they probably eat up another one-third of your electric bill. If you can tolerate a decrease in room temperature during the cold season and an increase during the warm season, you will save energy. Note that your air conditioner is the bigger lever of the two: when cooling a room, you must vent the hot air of the room into the even hotter air outside, and moving against entropy requires extra energy. An air-conditioning unit will require about double the electricity of a space heater to achieve the same degree change in temperature in a given room. Can you go without A/C for more days of the year? How much can you turn the heat down during winter? In theory (with a fireplace and a fan), you could all but eliminate this heating/cooling lever. Now you've cut your total electricity use by 60 percent.

The next biggest levers in your home are the objects that heat or cool small spaces: your clothes dryer, your stove, your dishwasher, and your refrigerator/freezer. Altogether they use about 15 percent of the total electricity in your home. Could you use them less often or run them at a different temperature? Admittedly, these are not strong levers, but every little bit counts.

Ironically, your television, your computer, and the lights that brighten your home—things you are probably careful to turn off when not using—contribute very little to your total electricity usage. You'd have to leave a sixty-watt light blazing 24/7 for a whole year in order to approach the amount of electricity that your electric stove consumes off-and-on. For slobs like me, the bad news is that this set of weak levers also includes your vacuum cleaner. If you vacuum your house once a month instead of once a week, you've barely shaved off one-third of 1 percent from your total electricity use for the year.

Nonetheless, by making all of these changes, you have now lowered your home electricity use by a whopping 70 percent. This lowers your tally from just over the average modern use in the OECD (ten megawatts per hour per year) to the 1965 average of Switzerland that I keep harping about. If each of the 1.3 billion people in the OECD made similar sacrifices, it would lower global electricity use by 25 percent and lower fossil fuel use and carbon dioxide emissions accordingly.

Maybe electricity use is not your thing; maybe you are more concerned with meat intake, or food waste, or car commuting, or airline travel, or pesticide use. Regardless of your mission, start in your own home and expand from there. I promise you'll be surprised at how far abroad it takes you.

The above may seem like an impossible task, but so did curing tuberculosis or putting a man on the moon or building the Great Wall across China or starting a new nation based on all men being equal or sailing across an unknown ocean in search of an unmapped land. History teaches us that these challenges were thoroughly overcome, via means honorable or shameful, despite the fact that they were originally deemed both ridiculous and impossible.

In most ways, we are just as noble and frail and flawed and ingenious as the people who cured and dared and built and forged centuries ago. Like them, we are ultimately endowed with only four resources: the earth, the ocean, the sky, and each other. If we can refrain from overestimating our likelihood of failure, then neither must we underestimate our capacity for success.

III. An Environmental Catechism

When you can measure what you are speaking about, and express it in numbers, you know something about it.
—Lord Kelvin (1883)

GLOBALLY, since 1969 . . .

. . . population has doubled.

. . . child mortality has decreased by half.

. . . average life expectancy has increased by twelve years.

. . . forty-seven cities have grown to contain more than ten million people.

. . . production of cereal grains has tripled.

. . . the amount of crops that can be harvested per acre has more than doubled.

. . . the amount of land cultivated for farming has increased by 10 percent.

. . . production of meat has tripled.

. . . annual slaughter of pigs has tripled; chicken has increased
   sixfold; cattle has increased by half.

. . . consumption of seafood has tripled.

. . . the amount of fish taken from the ocean has doubled.

. . . the invention of aquaculture made fish farming the source
   of half of all seafood eaten today.

. . . production of seaweed has increased tenfold; we eat half of
   the total as hydrocolloid food additives.

. . . consumption of table sugar has nearly tripled.

. . . the amount of human waste produced each day has more
   than doubled.

. . . waste of edible food has increased such that it now equals
   the amount of food needed to adequately feed all of the
   undernourished people on Earth.

. . . the total amount of energy that people use every day has
   tripled.

. . . the total amount of electricity that people use every day
   has quadrupled.

. . . 20 percent of the world's population has come to use half
   of the world's electricity.

. . . the total population living with no access to electricity has
   grown to one billion people.

. . . the number of airline passengers has increased tenfold,
   while the total distance traveled by rail has declined.

. . . the miles traveled by car each year has more than doubled;
   there are currently almost one billion motor vehicles on
   the planet.

. . . global fossil fuel use has nearly tripled.

. . . coal and oil use has doubled; natural gas use has tripled.

. . . the invention of biofuels has grown to consume 20 percent
   of the global annual harvest of grain.

. . . the production of plastic has increased tenfold.

. . . the invention of new plastics has grown to consume
10 percent of the fossil fuels used each year.

. . . the share of electricity generated using hydropower has
decreased to an all-time low of 15 percent.

. . . the share of electricity generated using nuclear power
increased to an all-time high of 6 percent.

. . . the adoption of wind- and solar-powered electricity
has grown to provide almost 5 percent of all electricity
generated each year.

. . . one trillion tons of carbon dioxide have been released into
the atmosphere from the burning of fossil fuels.

. . . the average global surface temperature has increased by
one degree Fahrenheit.

. . . the average global sea level has increased by four inches;
half of this rise was due to the addition of water melted
from mountain glaciers and polar ice.

. . . more than half of all amphibian, bird, and butterfly species
has declined in population; one-quarter of all fish and
plant species has declined in population.

IV. Sources and Suggested Reading

As yet a child, nor yet a fool to fame,
I lisp'd in numbers, for the numbers came.
—Alexander Pope (1734)

Upon finishing this book, readers may be asking themselves, Where in the world did she find quantitative data on feed conversion in domestic poultry, the frequency of adolescent hockey games in Manitoba, the volume of nocturnal urine production in humans, and the employment history of railway workers in Spain (and so much more)? Amazingly, it's all out there if you just look for it! While researching this book, I discovered the delightful truth that a database for pretty much *everything* exists, and the more you search, the better you get at searching. So before we're done, I want to share with you the sources that I used while researching and writing *The Story of More*, as well as some tips for pursuing your own research.

For many years, I have relied upon the Worldwatch Institute's publication *Vital Signs* (especially volumes 19, 20, 21, and 22) as a starting point for classroom discussions of the many issues presented in this book. These condensed guides to *The Trends That Are Shaping Our Future* are invaluable reading for anyone concerned about the future of our planet or curious about its past. Diverse organizations also consolidate climate change research for the public, such as Project Drawdown, a nonprofit coalition of scientists and entrepreneurs focused on the full suite of greenhouse gases, well beyond the basics of carbon dioxide that I presented here.

The numbers that I have offered, as well as the ones that I used to make calculations, were taken from several national and international public data sets and reports. I extensively used data archived by the many arms of the United Nations, including its Population Division (UNPD), Statistics Division, Departments of Economic and Social Affairs, Fisheries and Aquaculture, Human Settlements Program, Refugee Agency, and Global Happiness Council, as well as its Educational, Scientific and Cultural Organization (UNESCO), Food and Agriculture Organization, International Children's Emergency Fund (UNICEF), and World Health Organization (WHO). I also made use of the following UN reports: "Fish to 2030: Prospects for Fisheries and Aquaculture"; "2018 Revision of World Urbanization Prospects"; "Frequently Asked Questions on Climate Change and Disaster Displacement"; "Food Outlook" (2018); "OECD-FAO Agricultural Outlook 2018–2027"; and "A Guide to the Seaweed Industry."

I acquired additional data from several other international organizations, including the International Civil Aviation Organization, International Council for the Exploration of the Sea, International Energy Agency (IEA), Organisation for Economic Co-operation and Development (OECD), National Hydropower Association, Organization for the Development of Fisheries, Observatory of Economic Complexity (OEC), Organization of Motor Vehicle Manufacturers (OICA), International Starch Institute, and International Union of Railways. The IEA publishes a useful report titled "World Energy Outlook"; the Natural Resources Defense Council published another detailed report, "Wasted: How America Is Losing Up to 40 Percent of Its Food from Farm to Fork to Landfill" (2012). The World Bank provided data sets, as did the Global Energy Observatory and Gapminder.org. Four private companies supplied international data sets: the British Petroleum Global compilation on energy use; the Malcolm Dunstan and Associates database of concrete dams throughout the world; Plastics*Europe* (the Association of Plastics Manufacturers); and Elgiganten's database of energy efficiency for a multitude of home appliances.

The Intergovernmental Panel on Climate Change has published a series of assessment reports dated 1990, 1992 (supplement), 1995, 2001, 2007, and 2014, as well as a series of special reports on select topics including: 2000 ("Emissions Scenarios"), 2012 ("Renewable Energy Sources and Climate Change Mitigation"), 2012 ("Managing the Risks of Extreme Events and Disasters to Advance Climate Change Adaptation"), 2018 ("Global Warming of 1.5°C"), and 2019 ("Climate Change and Land" and "The Ocean and Cryosphere in a Changing Climate"). All of these were useful while researching this book.

Many of the examples in this book refer to the United States, the country of my birth. Data sets for these calculations were accessed from several branches of the United States Department of Agriculture (USDA), including the National Agricultural Statistics Service, Census of Agriculture, Economic Research Service, and Center for Nutrition Policy and Promotion, as well as Nationwide Food Surveys from 1935 to the present. Other U.S. federal entities that provided data were the National Census, National Park Service, Federal Register of the National Archives, Central Intelligence Agency World Factbook, Energy Information Administration (EIA), National Aeronautics and Space Administration (NASA), Geological Survey, National Agricultural Library, National Center for Health Statistics (NCHS), American Petroleum Institute, and the Department of State's Office of the Historian.

The National Academies of Science, Engineering and Medicine provided the following two relevant reports: "Resources and Man: A Study and Recommendations" (1969); and "Safety of Genetically Engineered Foods: Approaches to Assessing Unintended Health Effects" (2004). The Environmental Protection Agency (EPA) published a set of valuable reports that included "Pesticide Industry Sales and Usage: 2008–2012 Market Estimates" (2017); "Municipal Solid Waste Generation, Recycling, and Disposal in the United States: Facts and Figures for 2012"; "Light-Duty Automotive Technology, Carbon Dioxide Emissions, and Fuel Economy Trends: 1975 Through 2017"; "U.S. Households' Demand for Convenience Foods"; "Sugar and Sweetener Report" (1976); and "Sugar and Sweeteners Yearbook." I also used the USDA report "Family Food Consumption and Dietary Levels for Five Regions" (1941), and the Department of Health and Human Services' "2015–2020 Dietary Guidelines for Americans," 8th ed.

State- and city-specific data came from documents provided by the city of Philadelphia, Iowa State University, Michigan State Highway Department, Minnesota Historical Society, and St. Paul Public Works.

The country of Norway, where I now make my home, was very helpful in supplying me with access to the following databases: Statistical Institute, Fisheries and Coastal Affairs, Oil and Energy Program—all departments within the Norwegian Ministry of Trade, Industry and Fisheries. Three reports were also useful: the Norwegian Seafood Federation's "Aquaculture in Norway"; "Norway's Petroleum History," courtesy of norskpetroleum .no; the Norwegian Centre for Climate Services report "Sea Level Change for Norway: Past and Present Observations and Projections to 2100"; and the Norwegian Environment Agency report "The Impacts of 1.5°C: A Science Briefing."

Many of the claims made in this book were verified by news releases from PR Newswire, Reuters, and the USDA Newsroom. Several articles within periodicals were also used for this purpose. *The New York Times* was a preferred source of information, but the following sources also contributed: *The Atlantic,* Center for Science in the Public Interest, CNN, *Forbes,* Minneapolis *Star Tribune, National Geographic, National Review, Pacific Standard, Scientific American, Smithsonian, Vanity Fair,* and *The Washington Post.*

While researching *The Story of More,* I amassed a broad collection of the scientific literature, too many works to include here. However, I wish to note my reliance upon multiple works by the following scholars: Jonathan Bamber, Charles Benbrook, Mignon Duffy, Kerry Emanuel, John Harte, Ray Hilborn, Arjen Y. Hoekstra, Svetlana Jevrejeva, Matti Kummu, Tim Lenton, Diana Liverman, Katherine Meyer, Stuart Pimm, Barry Popkin, Roberto E. Reis, David B. Roy, William H. Schlesinger, Carl Schleussner, David Tilman, and David Vaughan. If I had to boil it down, under duress, to a recommendation of two works to be considered must-reads from the above, they would be the following:

1. New, Mark George, Diana Liverman, Heike Schroeder, and Kevin Anderson. "Four Degrees and Beyond: The Potential for a Global Temperature Increase of Four Degrees and Its Implications." *Philosophical Transactions of the Royal Society A: Mathematical, Physical and Engineering Sciences* 369, no. 1934 (2011): 6–19.

2. Vaughan, Naomi E., and Timothy M. Lenton. "A Review of Climate Geoengineering Proposals." *Climatic Change* 109, nos. 3–4 (2011): 745–90.

My fast-food study that I described within the appendix was published as follows:

Jahren, A. Hope, and Rebecca A. Kraft. "Carbon and Nitrogen Stable Isotopes in Fast Food: Signatures of Corn and Confinement." *Proceedings of the National Academy of Sciences of the United States of America* 105, no. 46 (2008): 17855–60.

For readers who may have caught the data bug, I want to recommend several aggregated resources where you can sift through international and national data on your own. I implore you not to put your research off, as there is no guarantee that these databases will exist next year, or next week, for that matter. Case in point: every two years since 2010, the EPA published a report titled "Climate Change Indicators in the United States." The 2010, 2012, 2014, and 2016 reports were invaluable public resources; they broke down the results of the state-of-the-art science clearly and accurately for lay readers, and illustrated trends using high-quality graphics. During the year 2018, however, no report was produced and no explanation was given for the omission. As far as I know, the EPA has no plans to release a report in 2020. As the leadership and priorities of our governments change, the type and amount of data that it chooses to make public may change also.

As of this writing, the World Bank maintains an open data site (data.worldbank.org) where you can visualize and download population, health, economy, education, and development data collected by the UNPD, UNICEF, WHO, UNESCO, IDS, OECD, and others. The Food and Agriculture Organization (FAO) of the United Nations maintains the FAOSTAT open data site (www.fao .org/faostat) where you can download files of agricultural data describing the production and consumption of crops, livestock, and other foodstuffs for every country of the world.

You may be able to find out what kind of power plant serves your neighborhood via the EIA's comprehensive site list-ing power plant schedules (www.eia.gov/electricity/data/eia923). British Petroleum Global maintains a database (www.bp.com/ en/global/corporate/energy-economics/statistical-review-of -world-energy.html) of the oil, coal, and gas use, production, and reserves, broken down by country, in addition to electricity use from both renewable and nonrenewable sources. For a compre-hensive picture of each country's annual imports and exports, I recommend the Observatory of Economic Complexity's website (oec.world/en/profile/country/usa/). The OECD's Data Depart-ment offers a site (data.oecd.org) where you can visualize and compare trends in passenger and freight transportation for each of the OECD countries and several others as well. The Organization of Motor Vehicle Manufacturers website (www.oica.net) offers downloadable data files describing production and sales statistics of cars and other vehicles all over the world.

The United States Department of Agriculture maintains a National Agricultural Statistics Service (quickstats.nass.usda.gov) where you can look up individual commodity yields for all plant and animal products produced in the United States, broken down by region, state, watershed, county, and zip code, as well as the Census of Agriculture website (www.nass.usda.gov/AgCensus/index.php), which offers data describing the size, status, and profit margin for the farms of America. The USDA also maintains a Foreign Agricultural Service (apps.fas.usda.gov/psdonline) describing individual commodities in terms of imports versus exports for every country in the world. The National Center for Health Statistics (under the umbrella of the Centers for Disease Control and Prevention) offers myriad data (www.cdc.gov/nchs/pressroom/calendar/pub_archive.htm) describing the food intake habits of Americans between 2009 and 2018, as well as directions for accessing older data sets. Finally, the United States Census Bureau performs population and demographic surveys of America that catalog housing patterns, salary by profession, commuting times, and so much more (www.census.gov/programs-surveys/decennial-census/decade.2010.html).

I made several choices when I wrote this book, and those choices had ramifications for the story that I've told. I mostly dealt with global data sets, as I wanted to discuss trends that were truly global in scale. Unfortunately, some key data sets are incomplete: for my numerous discussions of global energy, zero data exists for approximately twenty countries out of the world's two hundred or so. What these twenty countries have in common is first, they are located in sub-Saharan Africa and second, they are very poor countries with per capita income less than 10 percent of the global average. As such, these countries can be safely assumed to use low amounts of energy relative to the rest of the world (for now), but we should also remember that they contain two hundred and sixty million people in total—more than the population of Germany, France, Spain, and the United Kingdom combined.

There were many instances when I preferred to focus on the experience of women when relating the trends of the last fifty years. Usually my reasons were obvious: maternal mortality is an exclusively female statistic. Sometimes they were more complex: my analysis of the gender gap and how it might relate to population growth reflects both my examination of the data and my own experiences as a woman and mother.

I singled out several individual countries for analysis, namely, the United States for its outsized production and consumption of energy, corn, meat, sugar, and waste, Norway for the rapid evolution of its fishing industry, and Brazil for its ongoing loss of endemic species. The utility of dividing global data sets by country is rooted in the presumption that a sovereign nation is subject to a uniform economy and system of justice. For example, the country of Brazil is home to neither the greatest amount of forest (that would be Russia), nor the fastest shrinking forest (that would be Indonesia). However, Brazil *is* home to the greatest amount of shrinking forest in the world, all of it under the purview of a single government. Thus judicial or economic changes in Brazil have an outsized effect on global trends in forested land area, making Brazil a critical talking point within any discussion of global deforestation.

The trouble with dividing global data sets by country also lies in the presumption that a sovereign nation is subject to a uniform economy and system of justice. While researching chapter 8 ("Making Sugar"), my foray into data sets that specified black versus white Americans revealed significant disparities. For example, according to a USDA report from 1936, the households of "negro families" in southeastern states contained only 70 percent as much staple sugar as the households of "white farm operators" living in the same region. Seventy years later, in 2006, the National Center for Health Statistics reported that black adults were consuming twice as much sugar as white adults in the form of sugared beverages. Similarly, my general history of women's work in the United States diverged from that of women of color: while it is true that the percentage of all mothers who were employed outside the home doubled from 30 to 60 percent between 1976 and 1998, it is also true that the percentage of unmarried women of color (a number of them mothers) employed outside the home has not been lower than 60 percent for more than one hundred years. Clearly the general trends in American sugar consumption and female labor participation differ from the specific experience of Americans of color over the last several decades.

As the philosopher José Ortega y Gasset wrote in 1923, "To define is to exclude and negate." Any discussion of average trends effectively subsumes the experience of some, creating a picture that may be at odds with the experience of many. Analyses highlighting the history of individual countries and their subset populations will be the next step toward telling a fuller and more nuanced story than the one I presented here.

Finally, I wish to recognize the many farmers, foresters, ranchers, fishermen, nutritionists, businesspeople, factory workers, scientists, and engineers who read sections of this book and helped me clear up my use of the terminology, history, and specific workings of their particular fields. From this group, I wish to especially note the generous assistance of Cory Arjes, Elena Bennett, Clint Conrad, Brenda Davy, Mat Domeier, Andy Elby, Pierre Kleiber, Moses Milazzo, Matthew Miller, Paul Richard, Adolf Schmidt, and Reidar Trønnes.

Acknowledgments

My work on this book was encouraged and supported by many people, all of whom are very dear to me. I turned to Tina Bennett for guidance countless times while I was writing *The Story of More* and benefited enormously from her insight and experience. LuAnn Walther, my editor at Vintage, just plain *gets me* (it's an Iowa thing), and I am so grateful to her. Robin Desser and Ursula Doyle are women whom I have come to rely upon for thoughtful answers and careful eyes. Every author should have a trusted friend who reads her first drafts and gently steers her toward what she does best; for me that person is Svetlana Katz. Thirty-five years ago, Connie Luhmann and I sat side by side and learned grammar and shorthand from the same chalkboard. Who would have guessed that those two girls would grow up to be us, still diagramming sentences together, still close as sisters? Adrian Nicole LaBlanc continues to be generous in sharing her wisdom and experience with both writing and authorship.

There were two people who particularly encouraged me in this project but did not live to see it come to completion. Professor Fred Duennebier and the Reverend Fritz Fritschel steadily assured me of the importance and necessity of this book and of their faith in me as its author. My grief in losing them is tempered by my pride in following through. I am also indebted to each and every person—especially the good people at the Lutheran Church of Honolulu—who asked me when the next book was coming out. Your enthusiasm gave me permission to feed my fondest indulgence and call it "work," and I can never thank you enough. Last but not least, I want to thank whoever graffitied the electrical box at the corner of Blindernveien and Apalveien with: "We worship an invisible god and slaughter a visible nature—without realizing that this nature we slaughter is the invisible god we worship." It got me to thinking.

Also by Hope Jahren

Lab Girl is a book about work and about love, and the mountains that can be moved when those two things come together. It is told through Jahren's remarkable stories: about the discoveries she has made in her lab, as well as her struggle to get there; about her field trips – sometimes authorised, sometimes very much not – taking her from the American Midwest to the North Pole and from Ireland to Hawaii; and about her unswerving dedication to her life's work.

Visceral, intimate, gloriously candid and sometimes extremely funny, Jahren's descriptions of her work, her intense relationship with the plants, seeds and soil she studies and her insights into nature enliven every page of this thrilling book.

'Science is about a passion for ideas and the people who pursue those passions. Hope Jahren captures both . . . Engrossing'
Guardian

'Clear and compelling but also fiercely tender'
Rachel Cooke, *New Statesman*

'A fascinating account of plants, love and obsession'
Annabel Venning, *Mail on Sunday*

FLEET

To buy any of our books and to find out
more about Fleet, our authors and titles, as well
as events and book clubs, visit our website

www.littlebrown.co.uk

and follow us on Twitter

@FleetReads
@LittleBrownUK